MATHEMATICS RESEARCH DEVELOPMENTS

NETWORK THEORY AND ANALYSIS

Mathematics Research Developments

Additional books in this series can be found on Nova's website under the Series tab.

Additional e-books in this series can be found on Nova's website under the e-Books tab.

MATHEMATICS RESEARCH DEVELOPMENTS

NETWORK THEORY AND ANALYSIS

AMBER JORGENSEN
EDITOR

nova
science publishers
New York

Library of Congress Cataloging-in-Publication Data

ISBN: 978-1-53613-460-5

Published by Nova Science Publishers, Inc. † New York

Contents

PREFACE

In *Network Theory and Analysis*, the authors begin by examining the resource curse that explains the tendency of natural resource rich economies to perform poorly economically, as well as how these resources mediate other development challenges like poor governance, corruption, excessive borrowing, currency movement, neglect of manufacturing and agricultural sectors and violent conflict. They argue that the conceptualization of the curse suffers from methodological nationalism where state weaknesses and strengths are regarded as the main factors that condition the impact of resources. Next, a study focusing on Brazil's public health care is presented, suggesting that physicians with more professional connections probably take on more diverse professional occupations. Establishing policies like limiting weekly workload in the public health care system may lead to improving the efficiency of the overall network. Following this, the book examines social network analysis, the process of Identifying relationships between organizations, people, and connected information entities by mapping. The authors pay particular attention to the connection between actors, sense of independence, and emergent impacts. In the subsequent chapter, insight on IP addressing is provided, as well as on different IP addressing systems available today. The details of the prefix notation and CIDR notation are explained, and the various fields of the IPv4 header are discussed. Lastly, the various fields of the IPv6 and the difference between the IPv4 and the IPv6 addressing system are examined.

Chapter 1 - The resource curse explains the tendency of natural resource rich economies to perform poorly economically and how these resources mediate and are implicated in other development challenges like poor governance, corruption, excessive borrowing, currency movement, neglect of manufacturing and agricultural sectors and violent conflict. This chapter argues that the conceptualization of the curse suffers from methodological nationalism, where state weaknesses/strengths are regarded as the main factors that condition the impact of resources. This does not adequately account for diverse factors that shape development and geography. Based on actor network theory, the chapter posits that the curse is conditioned by a 'globalized assemblage' – interactions between and among national and local politics, transnational companies, technologies, actors, factors, structures and processes.

Chapter 2 - In the health literature, there is a paucity of studies that examine physicians' network of professional linkages in the public and private health sector. Therefore, there is a need to further investigate how the multiplicity of physicians' professional linkages shapes health network structures. The use of Social Network Analysis (SNA) tools in this context promises to be a fruitful line of inquiry with clear theoretical and practical implications. The study in a Brazil's public health care system is a unique context, because the lack of equipment, scarcity of resources for educational training and restrictive occupational licensing regulations causes the multiplicity of physicians' professional linkages. The results of this study presents some interesting conclusions in regard to the physician's professional linkages and their results in the physician's network structure. The overall size of the network structure is large, consisting of 789 actors. These results suggest that physicians with more professional linkages probably declare more different professional occupations. In regard to network centralisation and degree centrality, the results indicate that connections in this network are relatively centralized in one or more physicians, however the overall density is low, indicating that connections are somewhat redundant. Establishing policies, such as determination of limits to the number of professional linkages per physician and limits to weekly workload in the public health care system would lead to improve the efficiency of the overall network.

Chapter 3 - The availability of first hand information during emergency events helps in call for assistance and rescue. This chapter focuses on the assessment of such emergency events data extracted from the social network. For instance, tweets were extracted from twitter during the occurrence of any emergency events along with the various features. The features of tweets such as author details and the community surrounding are considered for the assessment. The most important features are identified using feature selection methods. The classification of tweets on the credibility is carried out using a feed forward neural network, which can reduce the loss on prediction and increase accuracy.

The tweets observed during emergency events have the ready to use information from the victims or people near the victims. These data, if properly analyzed, can give tremendous useful information for the emergency responders. The main challenge in the analysis is to know whether the tweet to be processed is credible or not. The solution for this problem is proposed using neural networks. Some of the problems of existing methodologies are not able to cover all the aspects of a tweet and minimal analysis of the features. The main advantage of the system lies in the learning stage where it uses neural network, the text is analyzed in many stages, requires less statistical training and the ability to detect complex relationships between variables. The over-fitting problem by incorporating a feature selection method is to extract the most relevant features. The tweets are extracted from the Twitter API. Three kinds of features can be constructed from the tweets namely Author, Tweet and Network Features.

Tweet Features: Length of the tweet, number of words, number of unique characters, number of hash tags, number of retweets, tweet is a retweet, number of special symbols [$, !], tweet is a reply, Number of @-mentions, number of retweets, has URL, number of URLs, use of URL shortened service *Author Features*: Registration age of the user, number of statuses, number of followers, number of friends, is a verified account, length of description, length of screen name, has URL, ratio of followers to followees. *Network Features:* Degree Centrality, Eigenvector Centrality, Betweenness Centrality, Kartz Centrality.

The important features that are useful for analysis are obtained using the Chi Square feature selection method. These features are considered for further feature construction process. The resulting data is fed to the Feed Forward Neural Network to assess the credibility into one of the two classes, Credible, Not Credible using graph theory.

Chapter 4 - This chapter provides an insight into an IP addressing, and different IP addressing schemes available today. Upon completing this chapter, the reader shall understand the details on IPv4 addressing, the basic concepts behind the IP addressing schemes such as Host ID, Network ID, Unicast addressing, Multicast addressing, Broadcast addressing, classifications of the IPv4 addressing schemes, Subnetwork, Subnet mask and default subnet mask. The details on the prefix notation and CIDR notation explained. Finally, discussed the various fields of the IPv4 header.

Chapter 5 - This chapter gives the introduction of the IPv6 addressing and the need for IPv6 addressing. This further discusses the IPv6 Addressing formats, the different address types, the special addresses in the IPv6. Finally covers the various fields of the IPv6 and the difference between the IPv4 and the IPv6 addressing scheme.

In: Network Theory and Analysis
Editor: Amber Jorgensen
ISBN: 978-1-53613-460-5

Chapter 1

POLITICAL ECONOMY TO GLOBALIZED ASSEMBLAGES: ACTOR NETWORK THEORY, HYDROCARBON ASSEMBLAGES, AND PROBLEMATIZING THE RESOURCE CURSE THESIS

Pius Siakwah*[*], *PhD

School of Business and Economics, Department of Tourism
University of Johannesburg, Johannesburg, South Africa

ABSTRACT

The resource curse explains the tendency of natural resource rich economies to perform poorly economically and how these resources mediate and are implicated in other development challenges like poor governance, corruption, excessive borrowing, currency movement, neglect of manufacturing and agricultural sectors and violent conflict. This chapter argues that the conceptualization of the curse suffers from methodological

[*] Corresponding Author: siakwahp@tcd.ie.

nationalism, where state weaknesses/strengths are regarded as the main factors that condition the impact of resources. This does not adequately account for diverse factors that shape development and geography. Based on actor network theory, the chapter posits that the curse is conditioned by a 'globalized assemblage' – interactions between and among national and local politics, transnational companies, technologies, actors, factors, structures and processes.

Keywords: relational, network, political economy, globalized assemblages

1. Introduction

Resource curse is the tendency for natural resource economies to perform poorly economically, and how natural resources mediate and are implicated in poor governance, corruption, excessive borrowing, currency movement, neglect/decline of manufacturing and agricultural sectors and violent conflict. Examination of the curse, however, over the years has focused on the national scale where the nature of the national political economy is viewed as the most crucial factor that mediates the opportunities or limits of resource-based development. Based on this, countries with institutions that guarantee accountability and eschew or limit corruption in the use of natural resource windfalls are probable to avoid the curse, while economies characterized by weak institutions, corruption and patronage will possibly manifest it (Alexeev and Conrad 2009; Brunnschweiler 2008; Brunnschweiler and Bulte 2008; Heum 2008). Resource rich countries like Norway, Australia, and Botswana however derive maximum benefits from their resources because their governments ensure accountability (Torvik 2009).

The above nationalist and teleological approach to explaining development where countries are supposed to imitate the 'best practices' of others (institutional practices) (Stiglitz 2006; Diamond 2005; Sachs 2005; Stiglitz and Charlton 2005) is challenging since national political economies differ. Sheppard (2011, p. 46) posited that even though such approach to analyzing development can be useful, a relational conception of the relationship between the economy, space/time and socio-nature within a broadly political economy, that creates space for theorizing development as an assemblage of possibilities, and locates actors in shifting alliances and rivalries is more enlightening. A relational perspective assists in appreciating development complexities spatially and temporally. Basing on the relational perspective, the chapter argues that most previous analysis of the resource curse 'over-territorialized' resource-based development challenges to national governments. Yet, a scrutiny of the hydrocarbon industry which produces the curse reveals complex 'globalized' assemblage, comprising states (oil importing and exporting ones), transnational companies, CSOs, technologies, capital, national and local actors, interacting in the construction and reconstruction and impacting the problematic impacts of natural resources (Siakwah 2017a, 2017b, 2017c, 2016). Based on a critical review of the curse literature, especially on the developing world, it is revealed that there globalized heterogeneous actors, and agencies that condition development outcomes.

This chapter examines how actor network theory (ANT) informed perspectives on networks, associations, relations and globalized assemblage explain and problematize the curse beyond the national scale, and how its manifestation is differentiated spatially and temporally. The rest of the chapter is organized into four sections. Section two revisits the curse literature and its problematic conceptualization–methodological nationalism. Actor network theory, associations and enrolment and how these aid 'deterritorialize or reterritorialize' the problematic impact of natural resources since the actors that mediate how resources are extracted and utilized go beyond the state are analyzed in section three. It examines relational assemblages, translations and relational networks and how they assist to elucidate the curse. In section four, the discussion focused on transnational hydrocarbon assemblage; comprising oil exporting and importing states, firms, CSOs and energy discourses and how interactions among these 'globalized' assemblages lead to production and spatial manifestation of the curse. It explores how assemblages enable us to understand how production and flow of resources go beyond the national economy since the global economy is a tangled web beyond geographical and political boundaries. The conclusion argues that the curse is conditioned by heterogeneous actors and structures, and manifest unevenly across spatial, temporal and social class.

2. The Resource Curse Revisited

The curse is initially explained by the tendency of resources rich economies perform poorly economically (Stevens 2015; 2003; Auty 2001, 1998, 1997, 1993, 1990; Sachs and Warner 1999, 1997; 1995; Karl 1997). It has since been extended to include how natural resources mediate and are implicated in development problems such as governance, corruption, excessive borrowing, currency movement, neglect/decline of manufacturing and agriculture sectors, environmental degradation and violent conflict (Watts 2010, 2009, 2008, 2003; Alexeev and Conrad 2009; Brunnschweiler and Bulte 2008; Heum 2008; Sachs and Warner 2001). The connection between resource endowment and development has been under scrutiny since apart from a few countries, most resource rich economies experienced poor economic growth compared to the resource poor ones (Stevens 2015; Ploeg 2011; Ploeg and Venables 2011; Collier 2010, 2008). The resource rich economies also performed abysmally on development indicators like level of poverty, social services access (education and health) and environment quality. Gylfason (2001) argued that educational attainment in resource rich country are adversely hampered as people become locked-in low-skill natural-resource-based industries. Some development researchers described the inverse relationship between resource abundance and poor development as a 'paradox of plenty' (Ploeg and Venables 2011; Karl 2007, 2004, 1997) since traditionally, economists regarded natural resource endowment as advantageous for development (Rosser 2006; Wright and Czelusta 2004, 2002; Krueger 1980).

Before the 1980s, the orthodox assessment of natural resources as advantageous prevailed (Rosser 2006). For instance, the geographer, Norton Ginsburg argued that a country's ownership of substantial and diversified natural resource endowment should be favorable for an accelerated economic growth (Rosser 2006; Higgins 1968; Ginsburg 1957). Development theorist, Rostow (1960) in the 60s also noted that natural resource endowments should facilitate developing countries' transition from underdevelopment to industrialization (Rosser 2006). Further, neoliberal economist, Krueger (1980, p. 2), in the 1980s advanced a related argument, noting that natural resources should facilitate industrial development by providing funds for investment in the broader national economy (Sheppard et al. 2009; Rosser 2006). Only a few economists, such as Prebisch (1950), Singer (1950) and Hirschman (1958) prior to the 1980s challenged the idea that natural resource abundance alone is necessarily sufficient for economic and industrial development. This is because the structure of the global economy, characterized by declining terms of trade for primary commodities relative to manufactured goods, commodity price fluctuations, and limited/lack of linkages between resource extraction and national economies positions resource exporting countries in dependency trap (Rosser 2006; Bulte et al. 2005). Declining terms of trade for commodities and price volatility make these countries to be governed via a transnational contract of extraversion – overly dependent on external actors for revenues (Carmody 2009).

Dimensions of the curse or overly dependent on natural resources can manifest in the form of neglect and/or decline in agricultural and manufacturing sectors and appreciation of local currency – Dutch Disease (Collier and Goderis 2007; Humphreys et al. 2007; Corden and Neary 1982). The Dutch Disease is named after the Netherlands' experience in the 1960s where its currency appreciated due to the influx of windfalls from gas, making imports cheaper, resulting in decline of manufacturing and agriculture (Ross 2012; Mcsherry 2006; Stevens 2003; Karl 2003). Natural resource rich economies in the developing world were under scrutiny because since the 1970s they appeared to perform poorly economically compared to resource poor economies such as Korea and Singapore (Humphreys et al. 2007; Auty 2001; Sachs and Warner 2001). The oil and mineral rich countries in Africa and elsewhere mostly underperformed economically (Bradshaw 2014; Bridge and Le Billion 2013; Carmody 2011; Watts 2010; African Development Bank 2009; Shaxson 2008; Oliveira

2007). Recent high growth rate due to commodity booms has not translated into poverty reduction (Stevens 2015).

Humphreys et al. (2007) and Heal (2007) contended that the dilemma of natural resource affluence and development is evident in countries like Korea, Singapore and Taiwan, where despite diminutive natural resources, they made significant economic strides, while most of the people in oil rich-countries in Angola, Gabon and Nigeria wallow in poverty. In Nigeria, whereas between 1965 and 2000, the country received US$350 billion dollars oil revenue, its GDP growth rate was 1.3% per annum during the same period, with most of the citizens living on less than US$1.5 dollars per day (Sala-i-Martin and Subramanian 2003). Maass (2009) posited that oil windfalls in Nigeria are 'mismanaged' by the elites, resulting in to poverty and violent conflict. Equally, natural resource-based or infused conflicts ruined Sierra Leone and Angola (Collier 2010; Oliveira 2010). In some cases, the elites used natural resource rents to consolidate power and personal enrichment (Oliveira 2007; Ghazvinian 2007). In the 1990s, it is reported that copious quantities of natural resources windfalls have been a curse instead of a blessing for the populace since these countries have increased tendency of experiencing political instability, environmental degradation and violent conflict (Ross 2013, 2003; Collier 2010; Karl 2003; Auty 2001). Resource rich economies are also characterized by declining terms of trade for natural resources to manufactured goods, lack of diversification, rent-seeking and irrational behavior of the political elites such as spending on exorbitant projects (Torres et al. 2013). Regardless of varied acceptance of the curse thesis as put forward by Auty (2001, 1993), Sachs and Warner (2001, 1995), and its visibility in Nigeria and Angola (Stevens 2015; Chindo et al. 2014; Oliveira 2010, 2007; Watts 2010, 2009), it is a subject of debate since some resource endowed countries such as Norway and Australia have performed better (NORAD 2013; UNDP 2011; Mehlum et al. 2006; Rosser 2006). These countries are however, not overly natural resource dominated, making the resource curse a complex phenomenon.

Substantial scholarly literature has however challenged the conventional wisdom held on the inverse relations between natural resources and development (Ross 2013; 2012; Auty 2001; Sachs and Warner 2001; Gelb 1988). Some studies noted that a critical examination of the political economy of natural resource rich countries, especially oil, revealed that although their growth rates might be erratic, they are comparable to other economies (Ross 2012). To other sceptics, it is not abundance of natural resources that is problematic, instead, it is the over dependence on the rent from natural resource with volatile prices that negatively affects the economies (Alexeev and Conrad 2009; Brunnschweiler and Bulte 2008). Wright and Czelusta (2004) posited that the US, Australia and Sweden used their advanced technologies to exploit their resources to galvanize industrialization. Norway and Botswana also exploited their oil and diamond respectively to promote economic development due to strong institutions and prudent investments policies (Moss 2011; Moss and Young 2009). Thus, although revenues from natural resources can be erratic, they can be harnessed for development. Consequently, even though the curse thesis provides a general picture about the relationship between natural resource endowment and development, it is a tendency, not a law (Auty 1993). Additionally, the existing explanation of the curse has not sufficiently account for external political and economic environments that interact with national and local politics and actors to determine development outcomes. The curse is produced and reproduced through interactions between natural resources, social, economic and political actors within and without the state.

3. Actor Network Theory: Networks, Associations and Assemblages

Actor network theory is not easily defined, it comprised heterogonous perspectives put together to aid explain the world's complexities (Law 1999, 1986; Latour 1988; Callon 1986). As a concept, ANT can be traced to science and technology studies (STS) and sociology of scientific knowledge (SSK) in the works of Latour (1988) on Pasteur, Callon's (1986) sociological analysis of technology and Law's (1986) social engineering studies. The stances of ANT have been adopted and applied to aspects social science like organization and management studies (Alcadipani and Hassard 2010), natural resource governance (Bodin and Crona 2009) and environmental justice (Holifield 2009). Duc to a recognition that nonhuman actors play vital parts in shaping the world, and as a critique of humanistic geographies that emphasis human creativity and consciousness, ANT is imperative in explaining socio-economic and political phenomena (Lorimer 2009). ANT highlights the substantial of 'non-human actors' in explaining geographical phenomena, although ascribing agency of non-humans is contested (Lorimer 2009). Within the tenants of ANT, phenomena are examined in their networks, relations, associations and assemblage. Doolin and Lowe (2002, p. 72) argued that societies should be explained 'as constituted by heterogeneous collectivities of people ... together with technology, machines and objects'. It is the complex inter-relations among the heterogeneous actors that makes up society, and connectivity is basic in defining sociality. Development should be understood as performed and shaped by an assemblage of actors and processes.

According to Law (1999), observing phenomenon through network lens means social divisions such as large and small; agency and structure; nature and society; human and non-human; knowledge and power; context and content; materiality and sociality; activity and passivity' become inconsequential. These things are interwoven and intrinsically linked. Social realities are fluid, multi-layered and multifaceted. Realities or phenomena are not out there to be discovered by either social actors or natural laws, they are performative and multiple (Mol 1999), and enacted through diversity of practices and connectivities. In practice, there is no gap between natural and social, since for example, the activities of natural scientists are influenced and shaped by the social settings in which they work.

Fundamentally, ANT aims to lessen the strict binaries between purified categories such as 'local', 'national' and 'global' (Johannesson and Baerenholdt 2009). A priori categories limit our appreciation of the world, hence that need to focus on networks, associations and relations since an event in one place can trigger other events in other spaces. The global is not independent of the national or local since they are mutually constitutive.

Local events have global effects and vice versa. For instance, the 2010 'Arab Spring' started in Tunisia as local event, but spread to Libya, Egypt, and other Arab countries. The same Arab Spring triggered other events such as global oil price hikes due to the conflict that characterized it which affected oil supply to the global market. Thus, phenomena are frequently caught up in web, and their interactions within and without the network shape their nature (Law 2009).

ANT Researchers conceptualized a priori categories as problematic because they believe in practice-based perspective, and rely on the 'general symmetry' principle which suggests that due to the instability and complexity of phenomena, entities need to be explained in their complexity (Johannesson and Baerenholdt 2009; Latour 1988). Concept of generalized symmetry repudiates categories like economic/cultural, natural/social and global/local since these phenomena are interconnected and constitutive. It is an anti-essentialist perspective that does not subscribe to binaries like nature/society or agency/structure (Johannesson and Baerenholdt 2009; Cressman 2009; Crawford 2005; Latour 2005, 1999). In explaining social phenomenon, one should focus on heterogeneous relations, while striving to place less emphasis on societal/technical and human/non-human divide (Alcadipani and Hassard 2010). Separating phenomena into societal/technical reduces their complexities, and how they are meshed into each, and interact with other actors/spaces. It is interactions among actors that shape phenomenon, and action and agency are distributed across these entities (Latour 1999). But it should be noted that 'by themselves, things don't act' (Law 1994, p. 485), but agency is a property that emerges via interactions between people and objects (Stein 2001).

Johannesson and Baerenholdt (2009) noted that ANT traces practices and processes through which societies are assembled. To understand the phenomena that constitute the world requires identifying the agents and relations that made up the assemblage (Latour 2005). In assemblage, emphasis is not only placed on ideas/discourses, we examine the material objects as well since they also assist in understanding the socicty. Latour (2005, p. 12) however, argued this required following the actors, and learning about what constituted their collective existence. Following the actors in their assemblage and interconnectedness aid researchers and policy makers to recapture the social as unstable associations (Latour 2005, 1988).

Assemblages are made up of a network of heterogeneous elements, and these ensure social resilience. Network is the sum of various actors and relations (Johannesson and Baerenholdt 2009). In effect, the social and natural world are heterogeneous networks. And this relative heterogeneity of the networked actors ensures their durability (Doolin and Lowe 2002; Latour 1988) because the actor–networks is intimately bound up with material and technical (Doolin and Lowe 2002; Latour 1991). ANT scrutinizes the heterogeneous actors in their associations as it is through this analysis that one can understand the function of each actor in the network and association (Cressman 2009). As Cressman (2009, p. 4) argued, associations illustrate 'how networks come to be larger and more influential than others, how they come to be more durable through enrolling both social and material actors, and where power comes from and how it is exerted'. And since ANT highlights network, it is also referred to as the enrolment theory (Crawford 2005). Enrolment is the process through heterogeneous actors are recruited to form a network based on the interest of the actors or a problem.

Translation on the other hand is the process through which an actor recruits other actors into its network so that whatever the recruited actors do will assist to strengthen and further the interest of the actor that recruits them. According Callon (1981, p. 211) translation 'involves creating convergences and homologies by relating things that were previously different'. Whereas enrolment entails recruitment of heterogeneous actors into a network, translation is the process through the recruited actors are used to strengthen the recruiter's position or ideas. Translation entails making connections, alliances and establishing communication among varied actors (Brown 2002). It is the stronghold position that enables an actor to control and utilize other actors' activities to further his or her interest (Latour 1988). Through translation, actors that are related in a socio-technical network negotiate and delimitate margins of maneuver (Callon 1986, p. 203). Translations encompasses the movement of technological processes, how ideas and plans are turned into action, how people and institutions are shaped to believe in the same thing, and transform technology to meet specific goals (Cressman 2009). It is a process of 'belonging by assemblages' (Lee and Stenner 1999), where diverse actors form a new phenomenon, yield a new entity and become stronger or more impactful than the individual actors. Connecting this to the hydrocarbon industry, oil companies can shape and control production in developing countries because they have the technology and capital.

In development studies and geography, ANT gained prominent in the 1990s and it is used to analyze global city (Smith 2003), urban assemblages (Prytherch 2011) and social relations in economic geography (Johannesson and Baerenholdt 2009; Jones 2008). Sheppard (2011) contended that we need to understand development from the relational perspective where economic conditions are explained by place based attributes and their connectivity to other places. Based on the arguments of Sheppard (2011), Latour (2005) and others, ANT's network, association and assemblage concepts are critical in explaining the hydrocarbon industry and its impacts. Network enables us to appreciate how states, institutions, transnational companies, energy discourse, and localized politics interact to shape development outcomes. ANT is influential in analyzing networks in economic geographies. Network geographies enable us to appreciate how economies are constituted, mediated and acted upon by a network of socio-cultural and political institutions and relations (Johannesson and Baerenholdt 2009).

Network is persuasive in urban and economic geographies, especially in production and distribution linkages among regions. States are linked by natural resources to other nations and companies. This highlights how production networks, trade, and foreign direct investments (FDIs) connect countries beyond national boundaries. In other words, while what happens in a specific region or state are significant in shaping these spaces, other processes and events outside also come into play. For example, concerning corruption in the natural resources sector, although is often attributed to national political elites diverting windfalls for personal use, transnational companies are occasionally implicated (Watts 2010; Oliveira 2007). In Equatorial Guinea, oil companies and international banks facilitated the transfer of state funds into personal accounts of the president and allies in the US (Shaxson 2008). The traditional political economy approaches where analysis is focused on the state although are valuable is also problematic. ANT elucidates analysis by paying attention to networks of institutions, social, transnational relations and non-human actors that combine to shape and re-shape development outcomes. This approach helps to examine how socioeconomic linkages work to shape the spatial organization in the global economy. Relational concepts like network and assemblage are prominent in explaining phenomena (Rungi 2012; Ernst and Kim 2002). Networks assist us to appreciate that local events have global implications.

4. Globalized Hydrocarbon Assemblages and the Curse

The framing and analysis of the implications of natural resources based development or the curse have been state-centered since both proponents of resource windfalls as advantageous or otherwise are fixated on the national political economy (Bridge 2008; Swyngedouw 2007; Gunton 2003). National institutions' qualities, accountability and patronage are recognized as main factors that mediate resource windfalls usage and their impact (Torres et al. 2013; Rosser 2006; Bulte et al. 2005; Wright and Czelusta 2004). While the national political economy cannot be underestimated in conditioning how windfalls are applied to development, focusing solely on it is problematic. The natural resource impacts are conditioned and shaped by a 'globalized assemblage'–interactions among and within national economies, transnational companies, capital, technologies, ideas, CSOs, global initiatives and local actors/politics.

Bridge (2008) posited that analysis natural resources based development, characterized by a national-scale focus that pushes questions of transnational actors and their impact on development into the background is difficult. Gunton (2003) highlighted the challenges with resource-based development based on poor governance nationally–overestimating windfalls, excessive spending and debt. Swyngedouw (2007) raised questions about state efficiency and competency without interrogating divergences of interests of transnational corporations, global energy security concerns and individuals that influence the structure of production and distribution of resource windfalls, and their national impacts. The ostensibly sole national scale analysis and focus on internal state weaknesses or strength does not convey the whole picture. Indeed, even where windfalls are managed well nationally, with significant economic impact, this does not necessarily trickle-down to localities and the impact is spatially differentiated.

Analysis of resource-based development suffers from methodological nationalism – where national government failure and/or competency are assumed to determine the success or otherwise of development. The hydrocarbon industry however, that produces and shapes the expression of the resource curse can be defined as a 'globalized assemblage'. Global assemblages, Smith (2010, p. 10) indicates, comprises 'tangible configurations through which global forms of techno-science, economic rationalism and expert systems gain significance and shape'. Global assemblage affords a lens via which to interrogate changing forms of organization and collective existence in the face of persistent political and ethical reflection and technological change (ibid.). Thus, scrutiny of the curse needs to be placed within the assemblage where activities of TOCs, global energy security, state and local politics mediate national and local development. The assemblage, Bridge and Le Billion (2013) argue, comprise of states (oil producing and consuming states), transnational firms and energy discourses. The boundaries of the assemblage can be extended and challenged to include human rights, climate change and environmental challenges. It should however, be emphasized that given thc differential outcomes of natural resources across and within states, states influence how natural resources are utilized. A critical enquiry of how natural resources are exploited and appropriated shows that the national states are influenced and acted upon by external factors like price volatility and local political patronage that can compel the elites to use windfalls for consumption instead of using it for long-term capital investments.

Existing analysis of the curse also failed to account for geography–how the resource curse manifests at local and national scales differently. In fact, although national scale studies are extensive (Ross 2012; Auty 2001; Sachs and Warner; Gelb 1988), local ones are limited (Obi and Rustad 2011; Watts 2009, 2004). The Niger Delta, Nigeria has been the focus of extensive research due to the violence and pollution that characterized oil exploitation in the region (Obi and Rustad 2011; Watts 2010, 2005, 2004). Other localized analysis of the curse included Ukiwo (2011), Ahonsi (2011), Ako (2011), and Idemudia (2011). Most of these studies are based on Nigeria which is one of the visible cases of the curse in Africa.

Coe et al. (2008) suggested that a relational thinking in which production is regarded as organized via inter-firm networks beyond state boundaries, and time-space sensitive has been missing in the scrutiny of the extractive industry and its impacts. The politics that characterizes that extractive industry, especially the hydrocarbon assemblage can be situated within a global network. A global network approach to analyzing extractive industry views the impact of natural resource on development (regional or national) as a dynamic product of a complex interactions between territorialized relational and production networks within changing regional governance structures (Coe et al. 2004). Even where the oil industry is understood as an enclaved, it is impacted by global, national and local actors. Development of global networks, facilitated by globalization makes it possible for production to take place globally (Rungi 2012). Within this network, globalized networks become mechanisms of knowledge transfer and building of local capabilities in the economic space (Ernst and Kim 2002). Globalized assemblage idea on network geographies, distribution and consumption elucidate how exploitation, procession and consumption of resources have spatial impacts.

The global production network is an arena of contested social relations involving transnational corporations, states as regulators and producers, labor, consumers and CSOs which shape natural resource governance. It involves how values or benefits from production are distributed across spaces (Lee et al. 2008; Lee 2006; 2000; Lee et al. 2004). In economic geography, it is argued that there is a need to move beyond the production network to value chain that examines how windfalls of production are distributed across spaces and class (Lee et al. 2008; Lee 2006). According to Lee (2006, p. 413), economic geography entails a struggle over value produced, consumption and exchange through social construction of material value, capable of being sustained across time and space. Value is formed and circulated through the construction of circuits and networks (Lee 2006; 2002; Hudson 2005). Hence, economic geography is a complex social practice (Lee 2006).

Economic geographies are social constructs based on exchange of knowledge and production possibilities (Lee 2000). The economy is embedded in everyday human practices (Lee et al. 2008). Circuits of value reach out across and through, multi-scalar spaces in which the environmental, material and social practices involved in them take place (Lee 2006). There are however differences in the conceptualization of the economy and how values are circulated. Mitchell (2002, p. 4) advocated for the construction of 'the economy'– 'a self-contained, internally dynamic and statistically measurable sphere of social action, scientific analysis and political regulation'. But Lee (2006, p. 417) argued that the bounded jurisdictional spaces of governance in which regulatory practices are established and implemented–by the state system are interrupted and shaped by changing relational geographies of flows. Thinking spatially and relationally about economic geographies enabled a reconsideration of the economic as a category, and the nature of relations between the economic, social, political, cultural and environmental systems (Lee 2000; Massey 1999). Based on ANT, the social and economic are mutually constitutive (Lee et al. 2008), hence economic geographies are analyzed as set of social relations and material processes that are exploitative in the process of production, value extraction, exchange and consumption circuits (Lee et al. 2004).

Based on network and relational perspectives, the hydrocarbon industry is perceived as constituted, in part, of – relational geographies of value circulation within network of actors and institutions. It is this network space of governance in which production, regulation and appropriation of value are established and implemented by global, state and local actors, structures and agencies. Lee (2006, p. 420) argued that once specific social relations of value start, those that are engaged in and benefiting from the system have a vested interest in extending and sustaining it. Thus, to be successful, the actors should be capable of being able to sustain the social relation over a sufficient geographical space and time to facilitate the economic rationality which enables social life to proceed and reproduce (Lee 2006). The circuits of economic value involve activities that are unstructured and complex, and go beyond state governance (Lee 2006). It involved practices of transnational and local actors.

It is not only changes at national or regional governance that influence how natural resources shape development but also global governance changes. For instance, the Extractive Industries and Transparency Initiative (EITI) and Publish What You Pay (PWYP) are global initiatives that promote accountability and transparency in the extractive sectors (EITI 2013, 2011). Over the years, the focus on corruption associated with the curse has focused on national political elites, but in practice, it involves the activities of the TOCs who are sometimes not ready to disclose payments to national governments because of competition between firms for contracts. This lack of disclosure can enable the elite to 'mismanage' windfalls (Ghana Extractive Industries and Transparency Initiative (GHEITI) 2013). The global initiatives are however not binding or is there a global governance mechanism to enforce them.

The governance structure of the extractive industry is significant in appreciating Africa's integration into the global economy. Trading in natural resources such as oil, gold and diamond, and import of manufactured products form the backbone of trading relations between Africa and the world. Yet, because African countries do not determine the prices and terms of trade of their natural resources, this affects their capacity to get maximum benefits from these natural resources. According to Dicken (2011), since the 1950s, volatile aggregate economic growth and uneven interconnectedness characterized the global economy where developing economies import expensive manufactured products and export low-priced raw materials.

Notwithstanding the unfavorable global interconnectedness for African countries, states still offer the political structure within which regulation of transnational companies occur, including negotiation of tax regimes (Coe et al. 2008). Yet, due to the value of trade in the global economy, analysis of development needs to focus on how states are interconnected into systems through uneven production and consumption networks. Regions or states are not bounded, instead they are interrelated with diverse socially constructed scales of activities and spaces (Dicken 2011). As Mitchell (2000, p. 392) suggested, 'thinking in terms of networks forces us to theorize socioeconomic processes as intertwined and mutually constitutive.' The embeddedness of states within national boundaries is quizzed by ANT's assemblage approach which traces practices and processes of socioeconomic outcomes in a globalized space.

Dicken (2011), Jones (2008), Mitchell (2000) and others advocated for relational and network approaches to analyze social phenomena to avoid the tendency of reducing the problems associated with natural resources to only national challenges–methodological nationalism. An analysis of the curse revealed that, it is intertwined and enmeshed with global energy concerns and local politics. This complex includes–TOCs concerned with profit, developed and emerging economies worried with energy availability and affordability, oil exporting governments concerned with revenue, and CSOs that are engaged with governance, environment and socio-economic development issues. The curse is a product and function of globalized assemblages – interactions between global, national and local politics, actors and processes. Relations among and between the actors and processes in the globalized assemblage are characterized by collaboration and disagreement. Bridge and Le Billion (2013) noted that the oil politics is characterized by competition, conflict, and cooperation; and these define the social and geographical distribution of the 'goods' and 'bads' associated with oil. The extractive industry revealed a network of inter-firm and firm-state relations that link national companies, resource holding states and transnational companies in complex lateral/horizontal relations (Bridge 2008). Competition for natural resources is the 'New Scramble for Africa'

(Carmody 2011). These competitions shape the hydrocarbon industry – hence, the curse should be rescaled as a spatial index of competing interests and politics.

Bridge and Le Billion (2013) posited that the new emerging geopolitics of oil centers on acceptability, affordability, accessibility and availability. Acceptability is concerned with how to make oil more socially and environmentally acceptable through reducing the conflicts and environmental pollution that characterized its production. Affordability entailed making oil cheaper by reducing taxes on it for the public or through subsidies as in the case of Nigeria, improving the energy supply infrastructure and incomes. Accessibility is guaranteed through diversification of supply sources. Accessibility can be ensured through diplomacy, as in the case of China where oil contracts are tied to loans, or military force as used by the US sometimes, or treaties like the EU-Russia Energy Dialogue (Bridge and Le Billion 2013). Thus, the interests and actors that shape the hydrocarbon industry transcend national governments; hence the challenges associated natural resources or oil transcend the countries. National governments in resource rich economies sometimes operate in the resource sector via national companies and as regulator of environmental, safety, health and tax systems. Technological and capital-intensive nature of the extractive industry nevertheless designed government in Africa to rely on established transnational companies like Shell, BP and national companies from China and Brazil to exploit the resources. These established companies can negotiate better contracts due to their technical and capital advantage. This makes the oil industry, from exploration, development, extraction, processing, distribution and consumption a complex web of interconnected relations that can be detrimental to resource rich economies in Africa with inadequate technologies. The global energy dilemma is centered around security, affordability, equity and sustainability (Bradshaw 2014), and the competing interests between economic growth, energy security and sustainable development condition the hydrocarbon industry and its differentiated impacts – the resource curse.

Conclusion: ANT, Globalized Assemblages, and the Curse

Application of ANT in research and policy has been limited because it is perceived as abstract (Albuquerque et al. 2013; Ernstson 2013), and more importantly for seemingly equating agency to human and non-human actors equally (Nigel Thrift 2008). Notwithstanding these early challenges, the ANT theoretical frameworks have recently been applied to explain the spatial processes and actions that influence development (Heeks 2013; Faik et al. 2013). This heuristic perspective highlights the complexities of actors and practices in development process (Heeks 2013). This chapter, thus, argues networks, associations and enrolment of other actors to build a stronghold position to control and utilize others to further one's interest based on ANT offer valuable insights into analyzing the hydrocarbon industry as an assemblage. The assemblage comprises competing actors and structures – nations, institutions, TOCs, CSOs, and local politics and actors. The actors in the assemblage are continually wrestling for significant positions within and outside the network (Smith 2010) since one's ability to control strategic sites within the networks provides capacity to shape actions or ideas of others (Yearley 2005). Hence, focusing analysis of the resource curse at the national political economy independently other actors and structures is obstinate. Natural resource rich economies are acted on and constituted via actors, agencies and structures, interacting internally and externally. The relationship between natural resources and development reveals a multi-layered actors and processes, spanning – global, national and local. Natural resource windfalls also have differentiated spatial and temporal impacts. ANT is concerned with fluidity and complexity, hence, it illustrates the dimensions of the curse not as inevitable outcomes naturally associated with resource rich countries, instead as contingent globalized assemblage.

The challenges of natural resource rich economies can be conceptualized as a mode of governance within a globalized assemblage, where national elites exercise authority over resource windfalls for personal enrichment in partnership with TOCs (Carmody 2009). Rightly, Obi (2013) and Shivji (2009) raised questions on how the resource curse is constructed, reproduced, whose interests do it serves, since it is the context in which resources are extraction and windfalls appropriated that determine whether they will be a curse or a blessing. Generally, Africa and most of the rest of the developing economies are inserted into a global economy that facilitate methodical export of natural resources through Structural Adjustment Programs (SAPs), World Bank, International Monetary Fund (IMF) and World Trade Organization (WTO) conditionalities (Carmody 2009). Bayart (2000) previously noted that through a 'strategy of extraversion', African countries are conditioned to export natural resource, and to facilitate expatriation of profits by TOCs from their economies.

Due to colonial legacy, the economies of most natural resource countries in African are characterized by a lop-sided dependence on the export of raw materials, and the import of manufactured goods. Africa and developing countries are integrated into a global economy in ways that are generally unfavorable to them and guarantees structural dependence. GDP growth that has occurred in Africa in recent decades was overwhelmingly characterized by the deployment and inflow of capital-intensive investment into extraction and exportation of natural resources such as oil, with limited value addition. Problematically, most of Africa's growth is externally driven–a classic example of dependency development (Taylor 2016a; 2016b), with a lack of serious structural change in the continent's economy. The impact natural resources (oil) on development, including industrialization and structural transformation are embedded and interwoven with socio-economic and political conditions or structures at global, national, and local scales (Siakwah 2016). And there are distinct temporalities and spatialities to natural resource impacts as well.

Based on how the curse is produced and manifested in the developing world, especially in Africa, the missing link in its analyses is how the actors, agencies and structures that shape and mediate it transcends national politics – globalized assemblage. Thus, although the national political economy is imperative, other actors and structures play equal important roles. Existing explanations for the curse have not sufficiently accounted for how interactions among the globalized actors and structures condition the development. Focusing on national politics alone has over-territorialized the challenges and/or opportunities associated with natural resource-based development (Bridge 2008). Thinking in terms of network and relations assist to 'deterritorialize' or 'reterritorialize' resource-based development, recognizing actors and processes that mediate resource extraction and utilization beyond state actors. Ayelazuno (2013) noted that in analyzing the curse, there is the need to move beyond the state to a critical political economy approach that examines the global political economic foundations of the curse and uneven capitalist development that position the resource endowed countries as primary producers and exporters. It is partly this unequal power relations and structures in the global economy that conditions natural resource-rich countries' development (Adusah-Karikari 2015). The oil industry in Africa is dominated by TOCs, transnational and local elites, with most of the oil refineries located outside, resulting in limited linkages to national economies (Carmody 2009; Le Billon 2001). The resource (oil) curse is profound among the poor whose livelihoods are sometimes alienated and threatened by the political economy of capital, plunder and greed of the transnational elites (Obi 2013). Because oil powers the global economy, it creates conditions where such states are subject to external interests, and the rent is also captured by dominant states and foreign private interests (Karl 2007). Based on ANT and globalized assemblage, the chapter argued that in analyzing the problematic impact natural resources, one must be conscious of relationality, networks and connectivity, examining the globalized political economy – link between national economies, societies and livelihood assemblages. While national politics, institutions and actors shape how resource revenues are utilized, they operate in tandem with localized and transnational actors and processes. Thus, globalized assemblage, an interaction among states, global and local actors, agencies and existing politico-economic structures that condition the challenges associated with natural resource-based development. Development challenges are not pre-determined, instead, they are socially created through the interplay of globalized actors and structures.

Acknowledgment

The authors acknowledge comments from Prof Padraig Carmody, Trinity College Dublin.

Declaration of Conflicting Interests: The author declared no conflict of interest to the authorship and publication of this article.

Funding: The author received no financial support for this research.

References

Adusah-Karikari, A. (2015). Black gold in Ghana: changing livelihoods for women in communities affected by oil production. *Extractive Industries and Society*, 2: 24–32.

African Development Bank (2009). *Oil and gas in Africa*. Oxford: Oxford University Press.

Adonis, B. A. (2011). Capacity and governance deficits in the response to the Niger Delta crisis. In: C. Obi and S. A. Rustad, eds., *Oil and insurgency in the Niger Delta: managing the complex politics of petro-violence*. London: Zed Books, pp. 28–41.

Ako, R. (2011). The struggle for resource control and violence in the Niger Delta. In: C. Obi and S. A. Rustad, eds., *Oil and insurgency in the Niger Delta: managing the complex politics of petro-violence*. London: Zed Books, pp. 17–27.

Albuquerque, J. P., Cukierman, H. L., Marques, I. C. and Feitosa, P. H. F. (2013). Challenging the ontology of techno-scientific artefacts: actor-network theory in developing countries. *Centre for Development Informatics*, Institute for Development Policy and Management, Working Paper 7, pp. 1 – 23.

Alcadipani, A. and Hassard, J. (2010). Actor-network theory, organizations and critique: towards a politics of organizing. *Organization* 17 (4), pp. 419 – 435.

Alexeev, M. and Conrad, R. (2009). The elusive curse of oil. *Review of Economics and Statistics* 91 (3), pp. 586 – 598.

Auty, R. M. (2001). The political economy of resource-driven growth. *European Economic Review* 45, pp. 839 – 846.

Auty, R. M. (1998). *Resource abundance and economic development*. UNU - World Institute for Development Economics Research, Helsinki: Oxford University Press.

Auty, R. M. (1997). Natural resources, the state and development strategy. *Journal of International Development* 9, pp. 651 – 663.

Auty, R. M. (1993). *Sustaining development in mineral economies: the resource curse thesis*. London: Routledge.

Auty, R. M. (1990). *Resource-based industrialisation: sowing the oil in eight developing countries*. Oxford: Oxford University Press.

Ayelazuno, J. (2013). Oil wealth and the well-being of the subaltern classes in Sub-Saharan Africa: a critical analysis of the resource in Ghana. *Resources Policy* 40, pp. 66 – 73.

Bayart, J. F. (2000). Africa in the world: a history of extraversion. *African Affairs* 99 (395), pp. 217 – 267.

Billon, P. (2001). Angola's political economy of war: the role of oil and diamonds, 1975 – 2000', *Africa Affair* 100, pp. 55 – 80.

Bodin, O. and Crona, B. I. (2009). The role of social networks in natural resource governance: what relational patterns make a difference? *Global Environmental Change*, 19 (3), pp. 366 – 374.

Bradshaw, M. (2014). *Global energy dilemmas: energy security, globalization, and climate change*. Cambridge: Polity Press.

Bridge, G. (2008). Global production networks and the extractive sector: governing resource-based development. *Journal of Economic Geography* 8, pp. 389 – 419.

Bridge, G. and Le Billion, P. (2013). *Oil*. Cambridge: Polity Press.

Brown, S. D. (2002). Michel Serres: science, translation and the logic of the parasite. *Theory, Culture and Society* 19 (3), pp. 1 – 27.

Brunnschweiler, C. N. (2008). Cursing the blessings? Natural-resource abundance, institutions and economic growth. *World Development* 36 (3), pp. 399 – 419.

Brunnschweiler, C. and Bulte, E. (2008). The resource curse revisited and revised: a tale of paradoxes and red herrings. *Journal of Environmental Economics and Management* 55, pp. 248 – 264.

Bulte, E. H., Damania, R. and Deacon, R. T. (2005). Resource intensity, institutions and development. *World Development* 33, pp. 1029 – 1044.

Callon, M. (1999). Actor-network theory – the market test. In: J. Hassard and J. Law, eds., *Actor-network theory and after*. Oxford: Blackwell Publishers, pp. 181 – 195.

Callon, M. (1991). Techno-economic networks and irreversibility. In: J. Law, ed., *Sociology of monsters: essays on power, technology and domination*, London: Routledge, pp. 132 – 161.

Callon, M. (1986). Some elements of a sociology of translation: domestication of the scallops and the fishermen of St. Brieux Bay. In: J. Law, ed., *Power, action and belief: a new sociology of knowledge?*, Keele: Sociological Review Monograph, pp. 196 – 233.

Callon, M. (1981). Struggles and negotiations to define what is problematic and what is not: the socio-logic of translation. In: K. Knorr., R. Krohn and R. Whitley, eds., *The social process of scientific investigation*, Dordrecht, Holland: Reidel Publishing, pp. 197 – 219.

Carmody, P. (2011). *The new scramble for African*. Cambridge: Polity Press.

Carmody, P. (2009). Cruciform sovereignty, matrix governance and the scramble for Africa's oil: insights from Chad and Sudan. *Political Geography* 28, pp. 353 – 361.

Chindo, M., Naibbi, A. I., and Abdullahi, A. (2014). The Nigerian extractive economy and development. *Journal of Studies and Research in Human Geography* 8 (2), pp. 71 – 87.

Coe, N. M., Dicken, P. and Hess, M. (2008). Global production networks: realizing the potential. *Journal of Economic Geography* 8, pp. 271 – 295.

Coe, N. M., Hess, M., Yeung, H., Dicken, P., Henderson, J. (2004). Globalizing regional development: a global production networks perspective. *Transactions of the Institute of British Geographers* 29 (4), pp. 468 – 485.

Collier, P. (2010). *The plundered planet: how to reconcile prosperity with nature*. Oxford: Oxford University Press.

Collier, P. (2008). *The bottom billion: why the poorest countries are failing and what can be done about it*. Oxford: Oxford University Press.

Collier, P and Goderis, B. (2007). Commodity prices, growth, and the natural resource curse: reconciling a conundrum. *Centre for the Study of African Economies*, Department of Economics, University of Oxford.

Corden, W. M. and Neary, P. J. (1982). Booming sector and de-industrialization in a small open economy. *The Economic Journal* 92, pp. 825 – 848.

Crawford, C. (2005). Actor network theory. In: G. Ritzer, ed., *Encyclopedia of social theory*, Thousand Oaks: Sage, pp. 1 – 4.

Cressman, D. (2009). A brief overview of actor-network theory: punctualization, heterogeneous engineering and translation. *Centre for Policy Research on Science and Technology*, School of Communication, Simon Fraser University.

Diamond J (2005). *Collapse. How societies choose to fail or succeed*. New York: Penguin.

Dicken, P. (2011). *Global shift: mapping the changing contours of the world economy*. 6th ed. London: Sage.

Doolin, B. and Lowe, A. (2002). To reveal is to critique: actor-network theory and critical information systems research. *Journal of Information Technology* 17, pp. 69 – 78.

EITI (2013). The EITI Standards. *EITI International Secretariat*, Oslo, Norway.

EITI (2011). *Extracting data: an overview of EITI Reports, 2005 – 2011*. Available at https://eiti.org/sites/default/files/documents/2011-11-28_Extracting_Data.pdf (Accessed 20th May 2015).

Ernst, D. and Kim, L. (2002). Global production networks, knowledge diffusion, and local capability formation. *Research Policy* 31, pp. 1417 – 1429.

Ernstson, H. (2013). Re-translating nature in Post-Apartheid Cape Town: the material semiotics of people and plants at Bottom Road. *Centre for Development Informatics*, Institute for Development Policy and Management, Working Paper 4, pp. 1 – 37.

Faik, I., Thompson, M. and Geoff, W. (2013). Facing the dilemmas of development: understanding development action through actor-network theory. *Centre for Development Informatics*, University of Manchester, Working Paper 2, pp. 1 – 33.

Gelb, A. H. (1988) *Windfall gains: blessing or curse*? New York: Oxford University Press.

Ghazvinian, J. (2007). *Untapped; the scramble for Africa's oil*. New York: Harcourt, Inc.

GHEITI (2013). *Ghana extractive industries transparency initiative* (GHEITI) Report, 2013. Ministry of Finance and Economic Planning (MOFEP), Accra.

Ginsburg, N. (1957). Natural resources and economic development. *Annals of the Association of American Geographers* 47, pp. 197 – 212.

Gunton, T. (2003). Natural resources and regional development: an assessment of dependency and comparative advantage paradigms. *Economic Geography* 79 (1), pp. 67 – 94.

Gylfason, T. (2001). Natural resources, education and economic development. *European Economic Review* 45, pp. 847 – 859.

Heal, G. (2007). Are oil producers rich? In: M. Humphreys., J. Sachs., and J. E. Stiglitz, eds., *Escaping the resource curse*, New York: Columbia University Press.

Heeks, R. (2013). Development studies research and actor network theory. *Centre for Development Informatics*, Institute for Development Policy and Management, Working Paper 1, pp. 1 – 29.

Heum, P. (2008). Local content development: experiences from oil and gas activities in Norway. *Institute for Research in Economics and Business Administration (SNF),* Bergen, February, Working Paper 2 (8), pp. 1 – 18.

Higgins, B. (1968). *Economic development: problems, principles, and policies*. New York: WW Norton Company.

Hirschman, A. (1958). *The strategy of economic development*. New Haven: Yale University Press.

Holifield, R. (2009). Actor-network theory as a critical approach to environmental justice: a case against synthesis with urban political ecology. *Antipode* 41 (4), pp. 637 – 658.

Hudson, R. (2005). *Economic geographies: circuits, flows and spaces*. London: Sage.

Humphreys, M., Sachs, D. and Stiglitz, J. eds. (2007). *Escaping the resource curse*. New York: Columbia University Press.

Idemudia, U. (2011). Corporate social responsibility and the Niger Delta conflict: issues and prospects. In: C. Obi and S. A. Rustad, eds., *Oil and insurgency in the Niger Delta: managing the complex politics of petro-*violence. London: Zed Books, pp. 167 – 183.

Johannesson, G. T. and Baerenholdt, J. O. (2009). Actor-network theory/network geographies. *International Encyclopedia of Human Geography*, pp. 15 – 19.

Jones, A. (2008). Beyond embeddedness: economic practices and the invisible dimensions of transnational business activity. *Progress in Human Geography* 32 (1), pp. 71– 88.

Karl, T. L. (2007). Ensuring fairness: the case for a transparent fiscal social contract. In: M. Humphreys., J. D. Sachs., and J. E. Stiglitz, eds., *Escaping the resource curse*, New York: Columbia University Press.

Karl T. L. (2004). Oil-led development: social, political, and economic consequences. *Encyclopedia of Energy* 4, pp. 661 – 672.

Karl, T. L. (2003). Bottom of the barrel: Africa's oil boom and the poor. *Catholic Relief Services*, June, 2003.

Karl, T. L. (1997). *The paradox of plenty: oil booms and petro-States*. California: University of California Press.

Krueger, A. (1980). Trade policy as an input to development. *American Economic Review* 70 (2), pp. 288 – 92.

Latour, B. (2005). *Reassembling the social: an introduction to actor network theory*. Oxford: Oxford University Press.

Latour, B. (1999). On recalling ANT. In: J. Law and J. Hassard, eds., *Actor network and after*. Oxford: Blackwell.

Latour, B. (1991). Society is technology made durable. In: Law, J. ed., *A sociology of monsters, essays on power, technology and domination*. London: Routledge.

Latour, B. (1988). *The pasteurization of France*. Cambridge: Harvard University Press.

Law, J. (2011). Heterogeneous engineering and tinkering. *Centre for Research on Socio-Cultural Change* (CRESC), Faculty of Social Sciences, The Open University, Walton Hall, pp. 1 – 18.

Law, J. (2009). Actor network theory and material semiotics. In: B. S. Turner, ed., *The new Blackwell companion to social theory*. Oxford: Blackwell Publishers.

Law, J. (1999). After ANT: complexity, naming and topology. In: Law, J. and Hassard, J, 1999, eds., *Actor network theory and after*. Oxford: Blackwell Publishers, pp. 1 – 15.

Law, J. (1994). *Organizing Modernity*. Oxford: Blackwell Publishers.

Law, J. (1987). Technology and heterogeneous engineering: the case of the Portuguese expansion. In: W.E Bijker., T. P. Hughes, and T. Pinch, eds., *The social construction of technical systems: new directions in the sociology and history of technology*. Cambridge, Mass: MIT Press.

Law, J. (ed.). (1986). *Power, action and belief: a new sociology of knowledge*? Keele: Keele Sociological Review Monograph.

Lee, N. and Stenner, P. (1999). Who pays? Can we pay them back?' In: J. Law and L. Hassard, eds., *Actor network theory and after*. Oxford: Blackwell Publishers.

Lee, R. (2006). The ordinary economy: tangled up in values and geography. *Transactions of the Institute of British Geographers* 31 (4), pp. 413 – 432.

Lee, R. (2002). Nice maps. Shame about the theory? Thinking geographically about the economic. *Progress in Human Geography* 26, pp. 333 – 55.

Lee, R. (2000). Shelter from the storm? Geographies of regard in the worlds of horticultural consumption and production. *Geoforum* 31, pp. 137 – 157.

Lee, R., Leyshon, A. and Smith, A. (2008). Rethinking economies/economic geographies. *Geoforum* 39: pp. 1111 – 1115.

Lee, R., Leyshon, A., Aldridge, T., Tooke, J., Williams, C. and Thrift, N. (2004). Making geographies and histories? Constructing local circuits of value. *Environment and Planning D: Society and Space* 22 (4), pp. 595 – 617.

Lorimer, J. (2009). Posthumanism/posthumanistic geographies. *International Encyclopedia of Human Geography*, pp. 344 – 354.

Maass, P. (2009). *Crude world; the violent twilight of oil*. London: Penguin Books.

Massey, D. (1999). On space and the city. In: Massey, D., Allen, J., Pile, S. eds., *City worlds open university*. London: Routledge, pp. 157 – 175.

Mcsherry, B. (2006). The political economy of oil in Equatorial Guinea. *African Studies Quarterly* 8 (3), pp. 23 – 45.

Mehlum, H., Moene, K., and Torvik, R. (2006). Institutions and the resource curse. *The Economic Journal* 116, pp. 1 – 20.

Mitchell, K. (2000). Networks of ethnicity. In: E. Sheppard and T. Barnes, eds., *A companion of economic geography*. Oxford: Blackwell Publishers.

Mitchell, T. (2002). *Rule of experts: Egypt, techno-politics, modernity*. Berkeley CA: University of California Press.

Mol, A. (1999). Ontological politics: a word and the same question. In: J. Law and J. Hassard, eds., *Actor network theory and after*. Oxford: Blackwell Publishers.

Moss, T. (2011). Oil to cash: fighting the resource curse through cash transfers. *Center for Global Development*, Washington D.C, January 2011, Working Paper 237.

Moss, T. and Young, L. (2009). Saving Ghana from its oil: the case for direct cash distribution. *Center for Global Development*, Washington D.C, October, Working Paper 186.

NORAD (2013). *Facing the resource curse: Norway's oil for development program*, Oslo.

Obi, C. (2013). Oil as the 'curse of conflict in Africa: smoke and mirrors. In: Abrahamasen, R. ed., *Conflict and security in Africa*. Suffolk, UK: James Currey, pp. 65 – 78.

Obi, C. and Rustad, S. A. (2011). Introduction: petro-violence in the Niger Delta – the complex politics of an insurgency. In: C. Obi and S. A. Rustad, eds., *Oil and insurgency in the Niger Delta: managing the complex politics of petro-violence*. London: Zed Books.

Oliveira, R. (2010). *Magnificent and beggar: Angola since the civil war*. London: Hurst and Company Publishers Ltd.

Oliveira, R. (2007). *Oil and politics in the Gulf of Guinea*. London: Hurst and Company Publishers Ltd.

Ploeg van der, F. (2011). Natural resources: curse or blessing? *Journal of Economic Literature*' 49 (2), pp. 366 – 420.

Ploeg van der, F. and Venables, A. J. (2011). Natural resource wealth: the challenge of managing a windfall. *Centre for Economic Policy Research*, December 2011, Discussion Paper 8694.

Ploeg van der, F. and Venables, A. J. (2009). Harnessing windfall revenues: optimal policies for resource-rich developing economies. *Center for Economic Studies* (CESifo), Working Paper 2571, pp. 1 – 36.

Prebisch, R. (1950). *The economic development of Latin America and its principal problems*. Lake Success, New York, United Nations.

Prytherch, D. L. (2011). Book review: urban assemblages: how actor-network theory changes urban studies. *Progress in Human Geography* 35 (6), pp. 855 – 856.

Ross, M. L. (2013). The politics of the resource curse: a review. In: C. and van de Walle, N. ed., *The handbook on the politics of development*. Lancaster, Oxford: Oxford University Press, pp. 1 – 29.

Ross, M. L. (2012). *The oil curse: how petroleum wealth shapes the development of nations*. Princeton: Princeton University Press.

Ross, M. L. (2003). Oil, drugs and diamonds: the varying roles of natural resources in civil war. In: Ballentine, K. and Sherman, J., ed., *The political economy of armed conflict*. Boulder, CO: Lynne Rienner, pp. 47 – 72.

Rosser, A. (2006). The political economy of the resource curse: a literature survey. *Institute of Development* (IDS), Working Paper 268, pp. 1 – 36.

Rostow, W. (1960). *The stages of economic growth: a non-communist manifesto*. Cambridge: Cambridge University Press

Rungi, A. (2012). Global value chains, state ownership and natural resources: EU policy on natural resources. *POLINARES* Working Paper, 47, April 2012.

Sachs, J. D. (2005). *The end of poverty: economic possibilities for our time*. New York: Penguin.

Sachs, J. D. and Warner, A. M. (2001). Natural resources and economic development the curse of natural resources', *European Economic Review* 45, pp. 827 – 838.

Sachs, J. D. and Warner, A. M. (1999). The big push, natural resource booms and growth. *Journal of Development Economics* 59, pp. 43 – 76.

Sachs, J. D. and Warner, A. M. (1997). Sources of slow growth in African economies. *Journal of African Economies* 6 (3), pp. 335 – 376.

Sachs, J. D. and Warner, A. M. (1995). Natural resource abundance and economic growth. *National Bureau of Economic Research*, Cambridge, MA, Working Paper 5398.

Sala-i-Martin, X. and Subramanian, A. (2012). Addressing the natural resource curse: an illustration from Nigeria. *Journal of African Economies* 22 (4), pp. 570 – 615.

Sala-i-Martin, X. and Subramanian, A. (2003). Addressing the natural resource curse: an illustration from Nigeria. *NBER* Working Paper Series 9804.

Shaxson, N. (2008). *Poisoned wells: the dirty politics of African oil*. 2nd ed. New York: Palgrave Macmillan.

Sheppard, E. (2011). Geography, nature, and the question of development. *Dialogues in Human Geography* 1 (1), pp. 46 – 75.

Sheppard, E., Potter, P. W., Faust, D.R, and Nagar, R. (2009). *A world of difference: encountering and contesting development*. New York: The Guildford Press.

Shivji, I. G. (2009). *Accumulation in an African periphery: a theoretical framework*. Dar es Salaam: Nkukina Nyota Publishers.

Siakwah, P. (2017a). Are Natural Resource Windfalls a Blessing or a Curse in Democratic Settings? Globalised Assemblages and the Problematic Impacts of Oil on Ghana's Development. *Resources Policy* 52: 122–133.

Siakwah, P. (2017b). 'Actor Network Theory, Globalised Assemblages and the Impact of Oil on Agriculture and Industry in Ghana', *Extractive Industry and Society*, 4: 462–472.

Siakwah, P. (2017c). Political Economy of the Resource Curse in Africa Revisited: The Curse as a Product and a Function of Globalised Hydrocarbon Assemblage. *Development and Society* 46, 1: 1–28.

Siakwah, P. (2016). *Are natural resources a blessing or a curse in a democratic setting - case study Ghana*? PhD Thesis, Trinity College Dublin.

Singer, H. (1950). The distribution of gains between investing and borrowing countries. *The American Economic Review* 40 (2), pp. 473 – 85.

Smith, J. (2010). *Biofuels and the globalization of risk: the biggest change in North-South relationships since colonialism*. London: Zed Books.

Smith, R. G. (2003). World city actor-networks. *Progress in Human Geography* 27 (1), pp. 25 – 44.

Stein, N. A. (2001). New directions in natural resource management: the offer of actor network theory. *IDS Bulletin* 32, 4:18 – 25.

Stevens, P. (2015). The resource curse revisited appendix: a literature review. *Energy, Environment and Resources*, pp. 1 – 42.

Stevens, P. (2003). Resource impact: curse or blessing? A literature survey. *Journal of Energy Literature* 9 (1), pp. 3 – 42.

Stiglitz, J. E. (2006). *Making globalization work*. New York: WW Norton.

Stiglitz, J. E. and Charlton A. (2005). *Fair trade for all: how trade can promote development*. New York: Oxford University Press.

Swyngedouw, E. (2007). Impossible/undesirable sustainability and the post-political condition. In: R. Krueger and Gibbs, D. eds., *The sustainable development Paradox*. New York: Guilford Press.

Taylor, I. (2016a). Dependency redux: why Africa is not rising. *Review of African Political Economy*, 43 (147), pp. 8 – 25.

Taylor, I. (2016b). Africa after the China boom. *Current History* 115 (781), pp. 193 – 196.

Thrift, N. (2008). *Non-representational theory: space/politics/affect*. London: Routledge.

Torres, N., Afonso, O. and Soares, I. (2013). A survey of literature on the resource curse: critical analysis of the main explanations, empirical tests and resource proxies. *Faculty of Economics, University of Porto*, Working Paper 2, pp. 1 – 24.

Torvik, R. (2009). Why do some resource-abundant countries succeed while others do not? *Oxford Review of Economic Policy* 25 (2), pp. 241 – 256.

Ukiwo, U. (2011). The Nigeria state, oil and the Niger Delta crisis. In: C. Obi and Rustad, S. A. eds.), *Oil and insurgency in the Niger Delta: managing the complex politics of petro-violence*. London*: Zed Books*.

UNDP (2011). *Managing natural resources for human development in low-income countries*, WP 2011 – 002, December 2011, Regional Office for Africa.

Watts, M. (2010). The rule of oil: petro-politics and the anatomy of an insurgency. *Journal of African Development* 11 (2), pp. 1 – 34.

Watts, M. (2009). Oil, development, and the politics of the bottom billion. *MacCalaster International* 24, pp. 79 – 130.

Watts, M. (2008). Imperial oil: the anatomy of a Nigerian oil insurgency. *Economies of Violence*, Working Paper 17, pp. 1 – 20.

Watts, M. (2005). Righteous oil? Human rights, the oil complex, and corporate social responsibility. *Annual Review Environmental Resources* 30, pp. 373 – 407.

Watts, M. (2004). Violent environments: petroleum conflict and the political ecology of rule in the Niger Delta, Nigeria. In: Peet, R. and Watts, M. eds., *Liberation ecologies: environment, development, and social movement*. Cornell: Cornell University Press.

Watts, M. (2003). Economies of violence: more oil, more blood. *Economic and Political Weekly* 38 (48), pp. 5089 – 5099.

Wright, G. and Czelusta, J. (2004). Why economies slow: the myth of the resource curse. *Challenge* 47 (2), pp. 6 – 38.

Wright, G. and Czelusta, J. (2002). *Resource-based economic growth, past and* present. Stanford University, Palo Alto.

Yearley, S. (2005). *Making sense of science: understanding the social study of science*. London: Sage.

BIOGRAPHICAL SKETCH

Dr. Pius Siakwah is a Research Fellow at the School of Business and Economics, Department of Tourism, University of Johannesburg, South Africa. He is a reviewer for a number of peer-review journals, including Geoforum, Resources Policy, Environment and Society, leading publishers in geographer and development studies. As an academic, Dr. Siakwah has published a number of articles in leading journals in geography, natural resources, development studies and network geographies. His PhD in geography was obtained from Trinity College Dublin (TCD), Dublin. He has a keen interest in political economy, natural resources, development geographies, network theories, and tourism and development.

In: Network Theory and Analysis
Editor: Amber Jorgensen
ISBN: 978-1-53613-460-5

Chapter 2

THE MULTIPLICITY OF PHYSICIANS' PROFESSIONAL LINKAGES IN THE BRAZILIAN HEALTH SYSTEM: A SOCIAL NETWORK ANALYSIS PERSPECTIVE

Simone Cristina de Oliveira Gobbo[1,*], PhD, José Alcides Gobbo Junior[1], PhD and Marcel Andreotti Musetti[2], PhD

[1]Department of Production Engineering,
São Paulo State University (UNESP), Bauru, SP, Brazil
[2] Department of Production Engineering,
University of São Paulo (USP), São Carlos, SP, Brazil

ABSTRACT

In health literature, there is a paucity of studies that examine physicians' network of professional linkages in the public and private

* Corresponding Author Email: sigobbo@sc.usp.br.

health sector. Therefore, there is a need to further investigate how the multiplicity of physicians' professional linkages shapes health network structures. The use of Social Network Analysis (SNA) tools in this context promises to be a fruitful line of inquiry with clear theoretical and practical implications. The study in a Brazil's public health care system is a unique context, because the lack of equipment, scarcity of resources for educational training and restrictive occupational licensing regulations causes the multiplicity of physicians' professional linkages. The results of this study presents some interesting conclusions in regard to the physician's professional linkages and their results in the physician's network structure. The overall size of the network structure is large, consisting of 789 actors. These results suggest that physicians with more professional linkages probably declare more different professional occupations. In regard to network centralisation and degree centrality, the results indicate that connections in this network are relatively centralized in one or more physicians, however the overall density is low, indicating that connections are somewhat redundant. Establishing policies, such as determination of limits to the number of professional linkages per physician and limits to weekly workload in the public health care system would lead to improve the efficiency of the overall network.

Keywords: social network analysis, Brazilian health system and physicians' professional linkages

INTRODUCTION

In recent years, social network theory has received attention from researchers and organisational managers due to its ability to analyse the patterns of interactions among the network "nodes" (Newman, 2003; Proulx et al., 2005). The theory has been applied to several areas of knowledge (Borgatti, 2005).

Increased interest in Social Network Analysis (SNA) in the health sector reflects the recognition that a web of social connections between patients and healthcare providers affects health processes and outcomes (Burt, 2014). SNA can be used to better understand communication flows and information exchange between actors (Autry & Griffis, 2005). Actors can be individuals or healthcare units (Carter et al., 2007).

Healthcare units are part of an intricate system of interconnected nodes that influence the dynamics of the network (Gobbo, 2010). Thus, the term 'network' is a word used extensively in health research and in the provision of health services and is used as a synonym for partnership, collaboration, alliance, and group. In other words, 'network' simply describes relationships between people, groups, or organisations. The physician's relationship with the patient must be multiform and reliable. Both parties should have some level of commitment to the relationship, and consequently, there is a need for intensive cooperation between different health units (Lämsä & Savolainen, 2003). In this sense, the interests of physicians are often related to the access of information and knowledge. Therefore, healthcare actors who position themselves at the core of a network are more likely to benefit from the dissemination of knowledge and to receive more information than peripheral actors (Simsek et al., 2003).

Social-professional networks, such as networks of physicians with professional links in public and private healthcare units, are under-researched, but are increasingly seen as crucial determinants in understanding and improving the healthcare network and patient-orientated services. Thus, there is a strong consensus among researchers in this area that it is important to develop models of evaluation of physicians' professional networks appropriate to the reality of the sector and with the support of data collection tools (Braithwaite et al., 2009).

In this research, the term 'physicians' professional network' means a professional that has a formal employment link to public and private healthcare units. In the literature (Maciel et al., 2010), the high workload and the multiplicity of professional links among physicians in the units that make up the healthcare networks determine the configuration of the network. This is important because little is known about how the multiplicity of physicians' professional linkages affects the configuration of the healthcare network.

The discussion about the analysis of social networks and their positive effects within the health sector is in the consolidation phase. In particular, most scientific works in the health sector, which analyse the effectiveness of networks, tend to be theoretical or conceptual, rather than empirical (Braithwaite et al., 2009).

However, there is still a lack of research that advances the analysis of social networks in the health system through the professional linkage of physicians; the number of studies in the public health sector is even smaller (Oliveira et al., 2004; Bittar & Magalhães, 2010; Bittar et al., 2011; Barata et al., 2010). Thus, there is a need to further investigate how the multiplicity of physicians' professional linkages shapes health network structures. The use of SNA tools in this context promises to be a fruitful line of inquiry with clear theoretical and practical implications.

The lack of equipment, scarcity of resources for educational training and restrictive occupational licensing regulations causes the multiplicity of physicians' professional linkages in Brazil to proliferate. This is a unique context and makes research on the multiplicity of physicians' professional linkages in Brazil desirable.

The public healthcare system in Brazil deals with increasing cost pressures to maintain quality of service (Barbeira et al., 2012). This is because through the creation of the Unified Health System (SUS), Brazil institutionalised the right to have healthcare (Viegas & Penna, 2013). The model adopted by the Brazilian public health system presupposes that health actions and services are organised in a regionalised and hierarchical healthcare network, composed of layers. Those layers are (1) Basic Attention Network (BA), the first level of healthcare. The population has access to basic specialties, such as medical clinic (general practice), paediatrics, obstetrics, and gynaecology; (2) Average Complexity (AC) are services such as hospital and outpatient consultations, exams, and some surgical procedures, and (3) High Complexity (HC) procedures that involve high technology and/or high cost, such as cardiac surgery and oncological treatments (Brazil, 2014).

Approximately 70% of Brazilian physicians have one or more positions in public and private health units, and 20% of these split their time between two or more cities (Ribas Filho et al., 2009). This decreases productivity since working hours are lost to travel between sites. The study of Brazilian physicians' professional linkages in public and private healthcare units can, therefore, provide valuable insights into knowledge dissemination and social influence within different health networks.

The following research questions were stated within this scope of research: RQ1 – Who are the central health actors in the physician's network with professional links in public and private health units? RQ2 - What are the centralisation and density indexes of the physician's network under study?

To address these questions, we developed an empirical quantitative study using SNA software. The measures used in this research were centrality and density. The research encompassed public health units located in Brazil. To the best of our knowledge, quantitative studies involving the analysis of social networks and the multiplicity of professional relationships of physicians with a focus on the health system are scare or non-existent. In addition, there is a lack of studies conducted in Latin America, and especially in Brazil.

This chapter contains five sections. The first one presents the purpose of the article and its justification; the second presents the concepts of Social Networks Analysis (SNA); the third characterises the object of study and describes the data collection and analysis procedures; the fourth covers our results and discussion, and the fifth provides main conclusions and highlight the contributions of the chapter.

2. Social Network Analysis (SNA)

Network theory involves the study of graphs as a representation of a symmetrical relationship, or, generally, of asymmetric relations between discrete objects. One of the applications of network theory includes social networks, that is, the social relationships between individuals. These actors are often people, but they can also be organisations or nations (Wasserman & Faust, 1994).

In this context, the literature presents two approaches: quantitative methods that use modelling, graphs, statistical tools, and organisational analysis (ONA) (Freeman, 1978; Barabasi et al., 2002; Chambers et al., 2012; Arroyabe et al., 2014; Schich et al., 2014; Zhao et al., 2013), and a qualitative aspect, which comes from sociology and mainly uses qualitative methods (Granovetter, 1973; 1985; Gilsing & Nooteboom, 2005; Gobbo, Junior, & Olsson, 2010; Gobbo et al., 2014). These two approaches are complementary and contribute to the analysis of networks.

In the qualitative analysis of the networks, there are structural and relational aspects that contribute to the concept of embeddedness, a term coined by Granovetter (1973) that means "degree of immersion." During periods of exchange, actors do not behave with perfect economic rationality because they are immersed (embedded) within social networks with other actors, who are able to provide greater access to resources and information (Granovetter, 1973; 1985).

In particular, social network theory is an appropriate basis for studying the outcomes for nodes in the network. With this in mind, there are two main social network constructs that differ in content (what flows through the ties), and the structure of their respective networks (density, centralisation, and node centrality) (Borgatti & Everetti, 2006).

For a better understanding of organisational networks, it is necessary to know their structure. The elements of this structure involve (1) centrality, which allows measurement of the involvement of an actor in the network; (2) structural equivalence, showing the actors that occupy similar positions; (3) density, which describes the overall level of connections between nodes in a network; and (4) cohesion, that one or more actors have power over the network (Wasserman & Faust, 1994). Networks can also be classified as dense or diffuse, and the greater the interconnection, the greater the density (Gnyawali & Madhavan, 2001). Structurally, networks may be dense or diffuse; relationally, networks may have strong or weak ties which are expressed through the nodes, links, or connections between the units of analysis (Sacomano Neto, 2004).

Networks as a closed system are usually composed of strong and dense ties between the actors. This type of network promotes trust and knowledge exchange, but the information is redundant. On the other hand, in networks with an open system, the degree of interconnection is relatively minor, involves mainly the transfer of resources and non-redundant information, and is composed of weak and dispersed links between their actors. In addition, the closed networks focus on social exchange trust and shared norms. The bonds that connect their actors are strong and dense (Granovetter, 1973; Coleman, 1988; Burt, 1992).

A key concept in analysing social networks is the centrality of the node. Centrality is the importance of a node due to its structural position in the network (Borgatti, 2009).

Borgatti (2005) determines centrality in the network according to the following factors: (1) the number of ties that involve a certain node; (2) to what extent the network revolves around a single node; (3) the extent to which the network resembles a star shape; (4) the difference between the score of the centrality of each node and the score of the most central node.

In this context, several authors (Freeman, 1978; Hansen et al., 2011; Scott, 2000; Borgatti, 2005) describe a set of centrality measures, such as: degree centrality, coreness, closeness; betweenness and eigenvector. In this research, we used centrality, betweenness, and density measures (neighbor and inclusiveness) because they are commonly used in similar studies and serve the purpose of this study.

Degree centrality measures the centrality of the network structure based on the degree of connections. The degree of centrality is calculated by the number of nodes that are adjacent to each node (Cyram, 2008) and is given by the following formula: centrality degree of node = $\frac{sum\ [\text{weight of incident connections}]}{\#\ node-1}$. The higher the degree of centralisation, the more central the actor is in the network (Britto, 2002). For Freeman (1978), a single central actor is equal to 1 and the other actors in the network are 0. Thus, the reference ranges from 0 (not central) to 1 (more central). This index helps measure the amount of direct involvement that a node has in the geodesic structure of the network (Borgatti & Everetti, 2006). The index also helps identify connected actors, popular actors, individuals who are likely to hold most information or individuals who can quickly connect with the larger network. Central actors can influence and be influenced directly (Borgatti, 2005). Degree centrality is a measure that captures the involvement and communication flows in a network. A node with a high degree of centrality occupies a structural position within a network that allows the node to maintain constant contacts with other network members and to mediate communication flows.

Betweenness is defined as the number of paths that pass through an actor between the shorter paths that connect two other actors (Borgatti, 2005; Baum et al., 2010). The betweenness centralisation index is a measure of the variability of the individual centrality betweenness scores. This index shows the centralisation score. Its value varies from 0 (non-central) to 1 (more central) (Cyram, 2008). Betweenness is a measure of control of information. High centrality of an actor's betweenness in the network indicates a stronger position in terms of brokerage, i.e., more control (and thus power) over the flow of knowledge through the network (Borgatti, 2005; Baum et al., 2010). In this sense, this measure can help in finding individuals who influence the knowledge flows around a system or who are able to disrupt the network; such as: key players, that if are removed, can cause disruption in the network (Borgatti, 2005). This helps to identify nodes, which have a large influence on the transfer of items through the network. In this study, the connections between the nodes of the network are considered undirected, i.e., the connections between nodes do not depend on their origin (in-degree equivalent to out-degree).

Neighbour, density, and inclusiveness allows analysis of the density of the network. The neighbour metric analyses degrees and types of nodes in the network. Two nodes are "adjacent" if there is a line between them. A node is an "incident" for a row if the node is one of the pairs of nodes that define the row. Degree nodal (connection) is the number of rows that are incident with it. It measures the size of its direct neighbourhood (Cyram, 2008). One of the most used concepts in network theory is density, which describes the general level of connections between nodes in a network. Density is measured by the ratio of the number of links present to the maximum number of possible links. A fully connected graph is one in which all points are adjacent, each node is directly connected to each other node in the network. Such completeness is rare, even in very small networks. The concept of density is an attempt to summarize the general distribution of links in order to measure how far the graph is from this state of completeness of connections. The more we are connected to each other, the denser this network will be. Density, then, depends on two other parameters of the network structure, these being the inclusiveness of the graph and the sum of degrees of its nodes. Inclusion refers to the number of nodes, which are included within the various connected parts of a graph. Put another way, the inclusiveness of a graph is the total number of points minus the number of isolated points (Scott, 2000).

3. Research Method

The selected research method involves processing quantitative data in order to map the network of physicians and analyse the centrality and density of that network. The study sample was limited to physicians with professional linkages in Brazilian public healthcare units of a Brazilian municipality.

The sample was defined from a Healthcare Regional Centre (Bauru Municipality) from São Paulo state. According to data from the National Register of Health Establishments (CNES), Bauru has 628 healthcare units (public and private). Seven of these are general hospitals, and three are specialised hospitals. Considered the most populated cities in the Centre-West area of São Paulo, Bauru has 364,562 inhabitants (Brazil, 2014) and is an important hub in the provision of medical services. The data sample included 65 public healthcare units from the Brazilian public healthcare System (Sistema Único de Saúde - SUS). Four healthcare units were excluded from this sample because the physicians who work in these healthcare units, do not have any ties with other healthcare units. To build our database, we collected data from the 61 healthcare units of the municipality of Bauru that were linked exclusively to the SUS. For each of unit, we listed all the physicians and accessed each physician's file. Therefore, 61 healthcare units, 789 physicians and 2,582 professional linkages in public health and private healthcare units were mapped.

We called the database DDB (Detailed Data Bank); it contained the following data: accessed healthcare unit, physician's name, Brazilian classification of occupation (BCO), healthcare units that have physicians' professional linkages, government and private spheres, state, workload, link in SUS and resident. In addition to the DDB, our qualitative data (SUS level of service, government and private spheres, number of physicians' linkages, among others) were prepared using spreadsheets in Excel and were later inserted into the NetMiner3 network analysis software.

4. Results and Discussion

The results of this study are presented in three sections: degree centrality, betweenness and density. In these sections, we describe and analyse the centrality and density measures obtained from mapping the network of healthcare units using the physicians' professional links.

4.1. Degree Centrality

This section presents the network centrality indicators by processing the data in NetMiner3 software. Table 1 presents the degree of network centrality and Table 2 presents the physicians' with the highest degree centrality.

Table 1. Distribution of the index of the degree of network centrality

Measure	Value *Centrality Degree*
Average	0.119
Standard deviation	0.149
Minimum	0
Maximum	0.669
Centralisation index of the network	55.029%

The results of Table 1 indicate that the network's average degree of centrality is 0.119, and the standard deviation is 0.149. In other words, node's connections seem to be dispersed. The minimum and maximum degree centrality, ranges between 0 and 0.669. The centralisation index is 55.029%. Centralization is a measure of how concentrated is the node's connection into one specific node. This indicates that connections are relatively centralized in one or more physicians.

The fifteen actors (physicians) with the highest centrality in the network are shown in Table 2. In terms of interactions with the actors in the network, the four most central physicians are notable. The physicians are the A223 and the A204 (both have 0.668 degree of centrality), followed by A251 (0.644 of centrality) and A172 (0.643 of centrality). It was observed that the number of physicians working at healthcare units is highly correlated with degree of centrality (correlation 0.802). Therefore, the higher the number of physicians at the healthcare unit, the higher will the degree centrality of that unit be in the sample.

Table 2. Physicians with highest degree centrality

Item	Physicians	Degree of Centrality
1	A223	0.668782
2	A204	0.668782
3	A251	0.64467
4	A172	0.643401
5	A197	0.639594
6	A215	0.629442
7	A143	0.623096
8	A220	0.615482
9	A208	0.615482
10	A198	0.615482
11	A173	0.615482
12	A148	0.615482
13	A145	0.615482
14	A222	0.612944
15	A194	0.612944

The most central physicians have the following occupations. The A223 has three occupations; they are clinical physician, anesthesiologist, and orthopaedist. The A204 physician is an anesthesiologist. The A251 physician has four stated occupations: clinical physician, anaesthesiologist, general surgeon, and urologist. Physician A172 has five occupations: clinical physician, gastroenterology, general surgeon physician, urologist, and radiologist.

It can be noted that these four mentioned physicians have public and private professional links and have the heaviest workload in hospitals. In addition, three of the physicians are anaesthesiologists and clinical physicians.

The central position of these actors allows them to access new knowledge (Borgatti, 2005), resources, and information in the network (Wasserman & Faust, 1994). Particularly in the health context, physicians who are located at the core of the network are more likely to benefit from the dissemination of knowledge and receive more information than peripheral actors.

Figure 1 illustrates the degree of centrality of the physicians in the network. The circle in this figure represents the amount of declared occupations that the physicians has in the network; therefore, the maximum number of occupations a medical professional can have in the study network is seven. The scale of grey circles demonstrates the physicians who have one to seven occupations. After analysing the number of occupations held by medical professionals in the study network, we found that the average number of positions held is two.

It is observed that six physicians in the network have the greatest number of occupations, and two of these stand out for their high degree of centrality. Of these six physicians, three of them presented seven different occupations. Physician A199 presents a high degree of centrality (0.604) in the network, and the other two physicians (A317 and A325) present low degrees of centrality in the network. The other three physicians who presented six different occupations are physician A203, who occupies a central position in the network (centrality of 0.579) and physicians A673 and the A151, who have low centrality in the network.

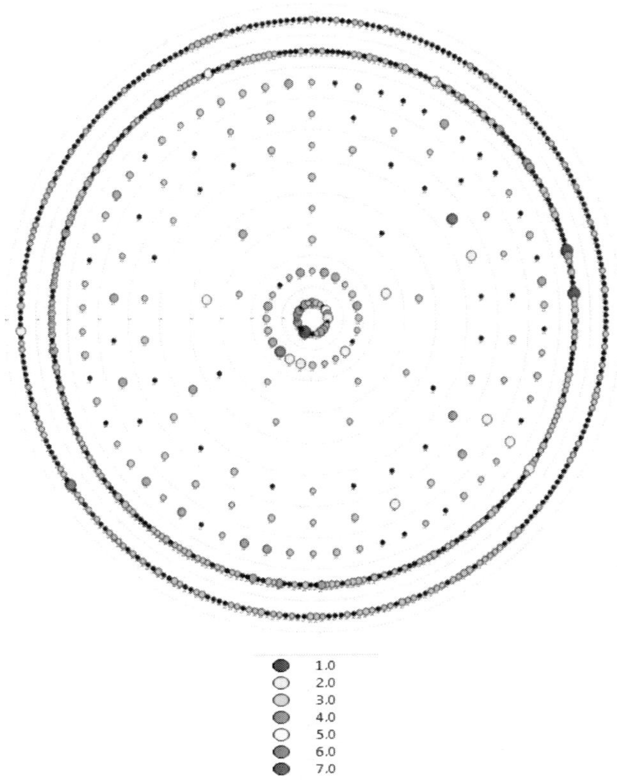

Figure 1. Centrality analysis (degree). Number of physicians' occupations. Concentric NetMiner 3 Preview.

The number of physicians who have one or two occupations is 627, and the number of physicians who have five or more occupations is 59 as shown in Graph 1.

The vast majority of physicians declare one or two occupations, however a small group declare five or more occupations. Eleven physicians have an unusual large number of professional linkages (11-27 professional linkages). It should be emphasised that the majority of occupations of these physicians' are in radiology and diagnostic imaging. The average number of professional linkages in the network is three. Why do a small group of physicians declare such a large number of occupations. This physicians' declaration of occupations is to support a large number of physicians' employments?

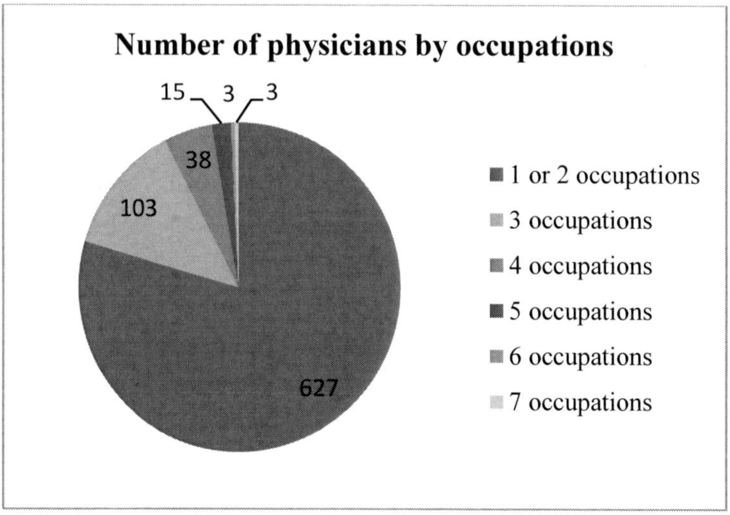

Graph 1. Number of occupations by physicians.

Figure 2 also illustrates the physician's degree of centrality in relation to the number of professional linkages in the network. The colour scale of the grey circle demonstrates the number of professional linkages of physicians, and varies between 1 to 27 professional linkages.

Among medical professionals with occupations in radiology, the A172 (centrality degree 0.643) is the only one that occupies a more central position in the network; the others with this occupation are located in the periphery of the network. This means that these physicians located on the periphery of the network are more likely to explore knowledge, probably through non-redundant contacts. Peripheral actors may be able to create new knowledge that can lead to the development of innovations (Gilsing et al., 2007). The number of professional links of the 15 physicians with the highest degree of centrality mentioned in Table 2 varies between 3 and 11 professional ties; for example, physician A223 has six links and A204 has four links, and both occupy the highest degree of centrality in the network. Therefore, a lot of professional links does not necessarily mean occupying a more central position in the network.

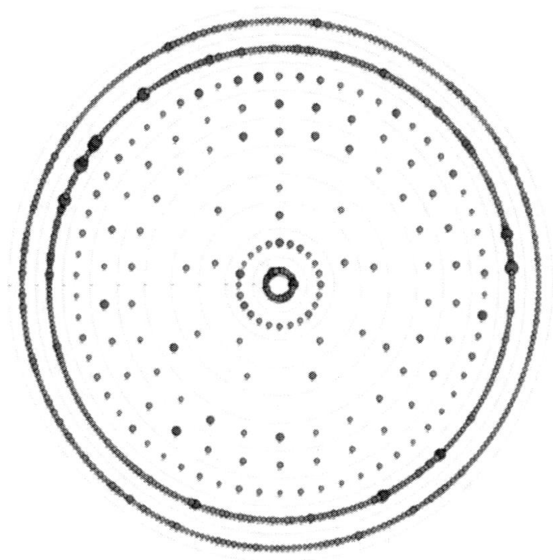

Figure 2. Centrality analysis (degree). Number of physician links. Concentric NetMiner 3 Preview.

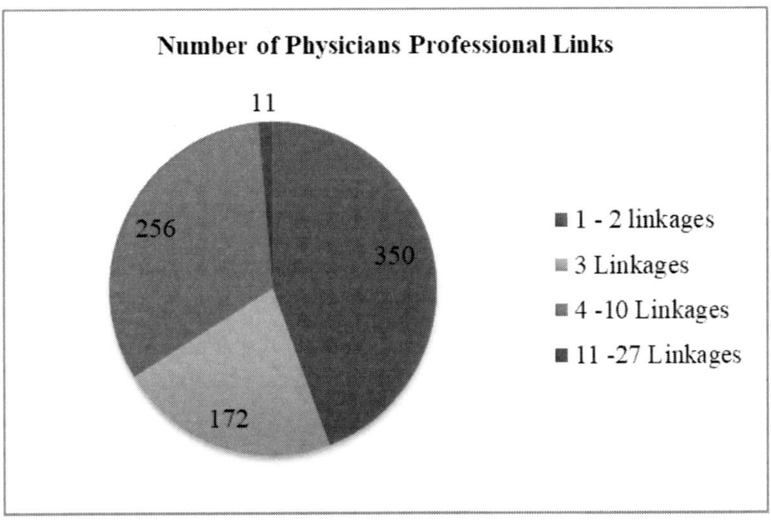

Graph 2. Number of professional linkages by physician.

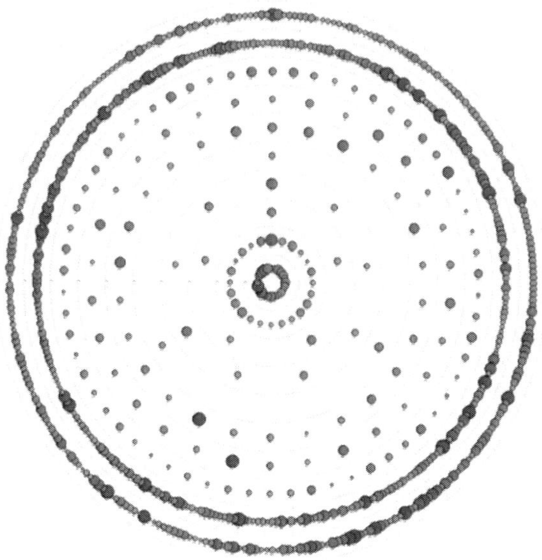

Figure 3. Centrality analysis (degree). Number of physician's weekly working hours. Concentric NetMiner 3 Preview.

Accordingly to Graph 2, the vast majority of physicians (522 physicians) declare between one or three professional linkages, however a large group of physicians (267 physicians) declare four or more professional linkages. Why do a small group of physicians build such a large number of employment linkages? Higher income, is an obvious response, however this physician's declaration of professional linkages would be linked to other reasons, such as knowledge access? This data seem to be supporting the expected multiplicity of physician's professional linkages.

Figure 3 shows the degree of centrality of the actors in the network and illustrates the network through the amount of weekly workload of the physicians. The scale of grey in Figure 3 varies between lower weekly hours and higher weekly hours. The average weekly workload of the physicians in the study network is 42 hours and ranges between 2 and 130 hours. It was found that physicians who have a high workload (100 and 130) are those who act as residents and have at least one occupation. Among all the occupations of the medical network, the three occupations with the highest average weekly working hours were resident physician (51 weekly hours), radiation therapist (40 hours a week), and family health physician (39 hours). It should be noted that the average was the total workload of each occupation of the network.

However, the four actors with the highest degree of centrality (represented by the red circles on the map) are above the average weekly hours. Those physicians are A204 (78 hours and 0.668 of degree centrality), A223 (56 hours and 0.668 of degree centrality), A251 (79 hours and 0.644 of degree centrality) and A172 (61 hours and 0.643 of degree centrality).

4.2. Betweenness

The Betweenneess node centrality quantifies the number of times a node intermediate flows between two other nodes (Table 3). The average node centrality is 0.001 and the Standard Deviation is 0.005. This means that the average degree of intermediation in the network is very low. The minimum betweenness centrality is 0 and the maximum is 0.092. The overall degree of betweeness centrality in the network is 0.091%. This means that the connections are somewhat redundant, with very few actors intermediating flows in the network.

Table 3. Centrality analysis (betweenness - node)

Measures	Value
Average	0.001
Standard Deviation	0.005
Minimum	0
Maximum	0.092
Centralisation Index (Betweenness)	0.091%

Table 4. Node Betweenness Analysis

Item	Physicians	*Betweenness Centrality*
1	A223	0.091766
2	A204	0.081119
3	A31	0.022641
4	A215	0.022301
5	A47	0.018177
6	A197	0.01777
7	A98	0.017546
8	A132	0.016523
9	A172	0.014981
10	A251	0.014392
11	A218	0.013098
12	A252	0.013028
13	A71	0.01178
14	A187	0.010489
15	A87	0.010485

Table 4 shows the 15 actors with the highest betweenness index. Physicians A223, A204, and A31 appear to be the three largest intermediaries in the network and perform most of the workload in hospitals. Their professional links are public and private. Both the A223 and A204 actors appear in the same position as they did in the centrality analysis (see Table 2). These three actors are noted as potential points for the control of communication (Fremann, 1979).

Figure 4 shows the degree of intermediation of the actors in the network of physicians (betweenness) with emphasis on the amount of weekly hours of physicians. The scale of grey varies between lower weekly hours and higher weekly hours. On the map, the physicians who are central in the network are in the best position to intermediate the information that circulates through the network; however, they are not those that present greater weekly workload. The most betweenness centrality actor is A223 (with betweenness degree 0.091 and workload 56 hours/week) represented on center of the map, but his weekly workload is above the average (42 hours) of the other members of the network.

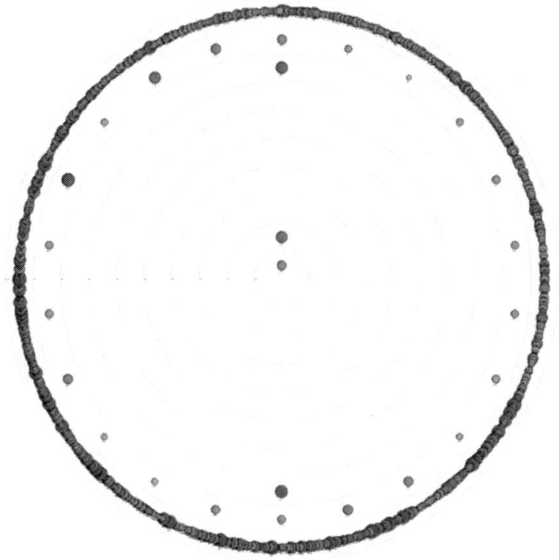

Figure 4. Betweenness centrality. Number of physicians' weekly working hours. Concentric NetMiner 3 Preview.

4.3. Neighbor, Density, and Inclusiveness

In the analysis of centrality (neighbor - degree), it is possible to identify the degrees and types of nodes in the network. Table 5 shows the results of neighbor degree centrality. The average nearest neighbor of a node is widely used to measure dependencies between neighbor nodes in a network. The sum of immediate neighbors in the network is 74102, with an, average of 93,919 neighbors for each node and with a standard deviation of 117.296. The minimum of neighbors for a node in the network is 0 and the maximum is 527. The number of isolated nodes is 9, which mean that they do not have any connections with other nodes, and, 14 dependent nodes, which are connected to only one node, exclusively. Inclusiveness is the number of connected nodes expressed as a number of total possible connections. Therefore, 98.859% represents that the vast majority of nodes are connected to peers (since only 9 physicians are disconnected from the others nodes in the network). The network density is low (0.119), that is, the number of lines present to the maximum possible. This means that the network density is low, however the inclusiveness index is high.

Table 5. Analysis of centrality (neighbor - degree)

Measures	Value *Neighbour – degree*
Sum	74102
Average	93.919
Standard Deviation	117.296
Minimum	0
Maximum	527
# isolated	9
# dependent	14
Inclusiveness (%)	98.859%
Network density	0.119

Table 6. Number of node type (neighbor – degree)

Isolated	Transmitter	Receiver	Loader	Ordinary
9	0	0	14	766

The results from Table 6 indicates 9 isolated (medical) nodes, 0 transmitter and receiver nodes, 14 loader and 766 ordinary nodes. The network under study is not directed, so the neighbor – degree analysis is pertinent with the fourth type of node, “loader” and requires that both input degree and output degree are equal to 1.

This section sought to map the network of physicians with professional links in public and private healthcare units, in addition to the overall network measures that were produced with the help of NetMiner3 software; in addition, basic descriptive statistics tools were used, such as average and correlation. We summarized some of variables, thought correlations, to make it clear some possible relations. The following variables were analysed:

1) Physicians' hourly workload and physicians professional linkages is 0,27, that is, a low positive linear correlation, if there were close to 1, there would be observed a strong linear correlation (Toledo & Ovalle, 1988);

2) The correlation between degree of centrality of physicians and total hours they worked was close to null (0.046); that is, correlations between variables range from small to almost imperceptible in significance. Although there is no correlation between these variables, it can be observed that the four physicians with the highest degree of centrality (see Table 2) have an above-average weekly workload (42 hours). Physician A223 worked 56 hours; A204 worked 78 hours per week; A251 worked 79 hours, and A172 worked 61 hours per week. However, the results show that most of the central actors do not have a higher hourly workload.

3) For degree of centrality of physicians and number of professional ties, the result was a positive linear correlation of 0.22. The four physicians with the highest degree of centrality present a number of above-average professional ties (three ties). A223 and A204 have six and four professional links, respectively. The third physician, A251, has 10 links and the medical room has 11 professional links. The central actors have more professional ties compared to the actors that are in the periphery of the network. Chart 1 summarizes the main discussions on the network measures and the main findings of the network of physicians.

4.4. Implications for Practice and Theory

This paper presents results relevant to the management of professional links of physicians. It was highlighted in Graph 2 (inserted in Section 4, subsection 4.1), that the number of professional linkages of physicians varies between 1 and 27. It is noticed that some actions regarding the management of medical professional bonds should be better disseminated throughout the public health system. The average number of physicians in the network is three, but about 33% of network physicians report having 4 to 10 professional links.

Chart 1. Summary of discussions on network measures at the analysis level - physicians

Analysis Level: Physicians		
Network Measures	Main Findings	Authors
Degree	The index of centralisation of the network is 55.029%, and the core of the network is composed of many actors. This index is composed of 242 actors and varies between 10% and 67% of centralisation. Of the 789 physicians, 781 physicians have some degree of centrality in the network. The central position of these actors allows them to access new knowledge, resources, and information in the network. It should be noted that the four most central physicians have public and private professional links, as well as the greatest workload in hospitals. Anaesthesiologist and clinical physician occupations are held by the three physicians occupying the most central position in the network.	Borgatti, 2005 Wasserman and Faust, 1994
Betweenness	The betweenness index is 0.091%. The three actors with the greatest betweenness spend most of their time in hospitals, and their professional ties are public and private. These actors may serve as a potential point for control of communication. These physicians have a stronger position in terms of brokerage, which is more control (and thus power) over the flow of knowledge across the network.	Fremann, 1979). Baum et al., 2010
Density	The network density is low (0.119). This density index means that 11% of the possible relationships within the network become effective. It promotes trust and cooperation among physicians and can generate social capital.	Coleman, 1988
Inclusiveness	The network inclusion index is 98.859%, the number of (medical) connected nodes expressed as a proportion of the total number of nodes. Since the index reaches almost 100%, it indicates that a large number of nodes are connected.	Zoppi, 2012

The multiplicity of physicians' professional linkages potentially leads to operational inefficiency in the health network, such as burnout syndrome and displacement among health units (Magalhães et al., 2015). Physicians seek to complement salary gains with various professional links (Maciel et al., 2010). Consequently, the weekly workloads of these physicians, who have accumulated various professional links, are higher. In the network of physicians under study, the average is 42 hours per week, but the range varies between 2 and 130 hours per week, which generates decreasing productivity, according to Pencavel's study (Smith, 2015). This illustrates the multiplicity of professional ties of physicians in different spheres of management.

Establishing policies to manage the professional links of physicians to improve the efficiency of the network and the health system is important. In the case of physicians' professional linkages, the determination of limits to the amount of professional linkages per physician working at a public healthcare unit and limits to weekly workload. This would also imply a reduction of labor costs for the overall network system, due to less frauds resulted from physicians working in a lower amount of weekly workload that the stated, and lower work benefits that need to be paid in different levels of the public health care system.

Establishing research questions would help to guide research in the field of physician's professional linkages. Therefore, we suggest the following questions, which would further be elaborated in research hypothesis. R.Q.1) Why do physicians engage in such a large number of professional occupations? R.Q.2) Does the large number of physicians' professional occupations support a large number of physician's professional linkages? R.Q. 3) Why physician's build such a large number of professional linkages? R.Q.4) The large number professional linkages would be linked to other reasons, such as knowledge access?

CONCLUSION

The results of this study presents some interesting conclusions in regard to the physician's professional linkages e their results in the physicians' network structure. The overall size of the network structure is large, consisting of 789 actors. The majority of physicians declare that have one or three professional linkages, however 34% of the physicians declare to have four or more professional linkages, ending in an astonishing number of 27 professional linkages. It would be rare to keep the same occupation for a physician working at so many different health care units, and physician reflects this in the number of occupations. The large number of physicians in the network declares one or two occupations, however a group of physicians (23% of the total) relate three or more different occupations. These results suggest that physicians with more professional linkages would declare more different professional occupations.

In regard to network centralisation and degree centrality, the results indicate that network the network's average degree of centrality is 0.119, and the centralisation index is 55.029%. Centralisation is measures of how concentrated are the node's connection into one specific node, therefore this indicates that connections in this network are relatively centralized in one or more physicians, however the overall density is low, indicating that connections are somewhat redundant. This result indicates that this network resembles a hub and spoke model health care network. Not every node is connected between them, however they connect to a central actor (which are physicians with professional linkages in very important hospitals and research centres).

The results allowed to the identification of the physicians typically found in the different levels of SUS care. Thus, in the center of the network are located the clinicians, general surgeons, and anesthesiologists, and in the periphery of the network are those with more specific specialties, such as: nuclear medicine physician, psychiatrist and physician of the family health strategy.

The results of this research contribute both to the theory, thought potential relationships, such as, the correlation between the number of physicians and unit's degree centrality and to practice, by the identification of the need for policies for the management of medical professional links in SUS. In addition, this study may serve as a pioneering study for academics and practitioners who decide to adopt this type of network analysis. The results of this research can contribute to the enrichment of knowledge about network mapping and the application of social network theory in the context of healthcare. In addition, this study can serve as a guide, both in the academic and the corporate sphere for this type of network analysis.

Because of the fact that the study was based in the Brazilian public healthcare system and limited to a local municipality (Bauru) of a regional health department, there are limitations of analysis and data collection. One limitation of this study was the temporal sample in data collection. Such as any networks, physician's networks are dynamic and changes during periods of time. Physicians working under the "More Physicians" programme (usually this group are foreign physicians and training physicians) are not included in the public database, so they were not included in this research.

REFERENCES

Arroyabe MF, Nieves A, Arroyabe JCF (2014). R&D partnerships: An exploratory approach to the role of structural variables in joint project performance. *Technol. Forecast. Soc. Change*, p. 1-11.

Autry CW, Griffis SE (2005). "A Social Anthropology of Logistics Research: Exploring Productivity and Collaboration in an Emerging Science," *Transportation Journal*, v. 44:4, pp. 27-43.

Barabasi AL, Jeong H, Néda Z, Ravasz E, Schubert A, Vicsek T (2002). Evolution of the social network of scientific collaborations. *Physica A*, v. 311, p. 590 – 614.

Barata LRB, Mendes JDV, Bittar OJNV (2010). Hospitais de ensino e o sistema Único de saúde. [Hospitals and the Unified Health System] *RAS*. v. 12, n. 46.

Barbeira M, Franco M, Haase H (2012). Knowledge sharing in networks within the healthcare sector: measuring knowledge management capabilities. In: *Proceedings of the 13th European Conference on Knowledge Management (ECKM)*, Spain, v. 1, p. 37-47.

Baum JAC, Cowan R E, Jonard N (2010) Network-Independent Partner Selection and the Evolution of Innovation Networks. *Management Science*, v. 56 (11), pp. 2094–2110.

Bittar OJN, Magalhães A (2010). Hospitais de Ensino no Estado de São Paulo : seis anos de acompanhamento. [Hospitals of Education in the State of São Paulo: six years of follow-up] *RAS*, especial, p. 155–213.

Bittar OJNV, Mendes JDV, Magalhães A. (2011) *Rede hospitalar no Estado de São Paulo: mapear para regular* [*Hospital network in the State of São Paulo: map to regular*]. São Paulo: SES/SP.

Borgatti S P, 2005. *Centrality*. Retrieved October 13, 2015, from http://www.analytictech.com/essex/lectures/centrality.pdf.

Borgatti SP, Martin G, Everett MG (2006) A Graph-theoretic perspective on centrality. *Social Networks*, v 28 p. 466–484.

Braithwaite J, Westbrook JI, Ranmuthugala G, Cunningham F, Plumb J, Wiley J, Ball D, Huckson S, Hughes C, Johnston B, Callen J, Creswick N, Georgiou A, Matibet, LB.; Debono D (2009) The development, design, testing, refinement, simulation and application of an evaluation framework for communities of practice and social-professional networks. *BMC Health Services Research,* 2009, 9:162.

Brasil. Ministério do Planejamento, Orçamento e Gestão (2014). Instituto Brasileiro de Geografia e Estatística. *Projeção da população do Brasil e das Unidades da Federação, 2014* [Ministry of Planning, Budget and Management (2014). Brazilian Institute of Geography and Statistics. *Projection of the population of Brazil and of the Federation Units, 2014*]. Retrieved August 06, 2014, from, http://www.ibge.gov.br/apps/populacao/projecao/.

Britto J (2002) Redes de cooperação entre empresas. In: Kupfer DE, Hasenclever L. *Economia Industrial: Fundamentos teóricos e práticas no Brasil* [Cooperation networks between companies. In: Kupfer DE, Hasenclever L. *Industrial Economics: Theoretical and Practical Foundations in Brazil*]. 9. ed. Rio de Janeiro, Editora Campus, 640p.

Burt RS (1992). *Structural Holes: The social structures of competition*. Harvard University Press, England.

Burt RS, Meltzer DO, Seid M, Borgert A.; Chung W, Colletti RB, Dellal G, Kahn SA, Kaplan HC, Peterson LE, Margolis P (2014). What's in a name generator? Choosing the right name generators for social network surveys in healthcare. *BMJ Qual Saf*, v. 21, p. 992–1000.

Carter CR, Leuschner R, Rogers DS (2007). A Social Network Analysis of the Journal of Supply Chain Management: Knowledge Generation, Knowledge Diffusion and Thought Leadership. *The Journal of Supply Chain Management*, v. 43, n. 2, p. 15–28.

Chambers D, Wilson P, Thompson C, Harden M., 2012. Social network analysis in healthcare settings: a systematic scoping review. *PloS One*, v. 7, n. 8.

Coleman JS (1988). Social capital in the creation of human capital. *American Journal of Economic Sociology*, 94 (Supplement), p. 95–119.

Cyram (2008) *NetMiner 3.3.0*. Seoul:Cyram Co., Ltd; 2008.

Freeman LC (1978) Centrality in social networks conceptual clarification. *Social Networks*. 1, 215-239.

Gilsing V, Nooteboom B (2005). Density and strength of ties in innovation networks: an analysis of multimedia and biotechnology. *European Management Review*, n. 2, pp. 179-197.

Gilsing VA, Lemmens, CEAV, Duysters G (2007). Strategic Alliance Networks and Innovation: A Deterministic and Voluntaristic View Combined, *Technology Analysis & Strategic Management*, 19:2, p. 227-249.

Gnyawali DR, Madhavan R (2001). Cooperative networks and competitive dynamics: A structural embeddedness perspective. *Academy of Management Review*, v. 26, n. 3, p. 431-445.

Gobbo SCO, Fusco JPA, Gobbo Junior JA (2014). An analysis of embeddedness in the value creation in interorganisational networks: an illustrative example in Brazil. *Int. J. Advanced Operations Management*, v. 6, n. 2, pp. 178-198.

Gobbo Junior JA, Olsson A (2010). The transformation between exploration and exploitation applied to inventors of packaging innovations. *Technovation*, n. 30, p. 322-331.

Gobbo SCO (2010). *Uma análise do embeddedness na formação de valor em redes de empresas: um exemplo de aplicação em uma rede do setor metal-mecânico.* [Dissertação de Mestrado]. [*An analysis of embeddedness in value formation in business networks: an example of application in a network of the metal-mechanical sector.* (Masters dissertation).] Bauru: Universidade Estadual Paulista, Campus Bauru.

Granovetter MS (1985). Economic Action and Social Structure: The problem of Embeddedness. *The American Journal of Sociology*, v. 91, nº 2, pp. 481-510.

Granovetter MS (1973). The strength of weak ties. *American Journal of Sociology*, n. 78, p. 1360-1380.

Hansen DL, Shneiderman B, Smith MA (2011). *Analysing social media networks with NodeXL: Insights from a connected world.* Morgan Kaufmann, Amsterdan. 284p.

Lämsä T, Savolainen T. Network Approach to Health Care Quality Improvement: the Case of a Finnish Hospital District, in E. Geisler, K. Krabbendam and R. Schuring (eds), *Technology, health care, and management in the hospital of future*, Praeger Publishers, Westport, pp. 3-12, 2003.

Maciel RH, Santos JBF, Sales TB, Alves MAA, Luna AP, Feitosa LB (2010). Multiplicidade de vínculos de médicos no Estado do Ceará. [Multiplicity of medical links in the State of Ceará.] *Rev Saúde Pública*, 44 (5):950-6.

Magalhães E, Oliveira ACMS, Govêia CS, Ladeira LCA, Queiroz DME, Vieira CV (2015). Prevalência de síndrome de *burnout* entre os anestesiologistas do Distrito Federal. [Prevalence of burnout syndrome among anesthesiologists in the Federal District.] *Rev Bras Anestesiol.* 65(2):104-110.

Newman M. E. J. (2003). The structure and function of complex networks. *Siam Rev.* 45, 167–256.

Oliveira EXG, Carvalho MS, Travassos C (2004). Territórios do Sistema Único de Saúde – mapeamento das redes de atenção hospitalar. [Territories of the Unified Health System - mapping of hospital care networks.] *Cad. Saúde Pública*, Rio de Janeiro, v. 20, n. 2, p. 386–402.

Proulx S. R., Promislow D. E. L., Phillips P. C. (2005). Network thinking in ecology and evolution. *Trends Ecol. Evol.* 20, 345–353.

Ribas Filho J M, Malafaia O, Czeczko NG (2009). The reality of Brazilian medical market. *Revista do Colégio Brasileiro de Cirurgiões*. Vol. 36, No. 3, Rio de Janeiro, July 2009.

Sacomano Neto M (2004). *Redes: Difusão de conhecimento e controle: Um estudo de caso na indústria Brasileira de caminhões*. (Tese de Doutorado) [*Networks: Knowledge diffusion and control: A case study in the Brazilian truck industry*. (Doctoral thesis)]. São Carlos: Universidade Federal de São Carlos, 2004.

Schich M, Song C, Ahn YY, Mirsky A, Martino M, Barabási AL, Helbing D (2014). A network framework of cultural history. *Science,* v. 345, p. 558-562, 2014.

Scott J (2000). *Social Network Analysis: A Handbook*, 2nd ed., Sage Publications, Thousand Oaks, CA.

Simsek Z, Lubatkin M, Floyd SW (2003). Inter-Firm Networks and Entrepreneurial Behavior: A Structural Embeddedness Perspective. *Journal of Management*, v. 29; pp. 427-442.

Toledo GL, Ovalle II (1988). *Estatística Básica.* [Basic Statistics.] São Paulo: Atlas, 1988.

Viegas SMF, Penna, CMM (2013). O SUS é universal, mas vivemos de cotas. [The SUS is universal, but we live on quotas.] *Ciência & Saúde Coletiva*, v. 18, n. 1, pp. 181-190.

Wasserman S, Faust K (1994). *Social Network Analysis: Methods and Applications*. Cambridge University Press, New York, NY.

Zhao JH, Zhou HJ, Liu YY (2013). Inducing effect on the percolation transition in complex networks. *Nature communications*, v. 4, p. 1-6.

Zoppi LG. Introdução ao NetMiner (2012). *Tree Branding. Apresentação ao Programa de Pós Graduação em Engenharia de Produção da UNESP* [*Tree Branding. Presentation to the Graduate Program in Production Engineering of UNESP*], Campus de Bauru, Bauru, 24 Ago 2012.

In: Network Theory and Analysis
Editor: Amber Jorgensen
ISBN: 978-1-53613-460-5

Chapter 3

SOCIAL NETWORK ANALYSIS USING GRAPH THEORY

V. Pattabiraman* and R. Parvathi*
School of Computing Science and Engineering,
Vellore Institute of Technology University, Chennai, India

ABSTRACT

The availability of first hand information during emergency events helps in call for assistance and rescue. This chapter focuses on the assessment of such emergency events data extracted from the social network. For instance, tweets were extracted from twitter during the occurrence of any emergency events along with the various features. The features of tweets such as author details and the community surrounding are considered for the assessment. The most important features are identified using feature selection methods. The classification of tweets on the credibility is carried out using a feed forward neural network, which can reduce the loss on prediction and increase accuracy.

The tweets observed during emergency events have the ready to use information from the victims or people near the victims. These data, if

* Corresponding Authors: pattabiraman.v@vit.ac.in, parvathi.r@vit.ac.in.

properly analyzed, can give tremendous useful information for the emergency responders. The main challenge in the analysis is to know whether the tweet to be processed is credible or not. The solution for this problem is proposed using neural networks. Some of the problems of existing methodologies are not able to cover all the aspects of a tweet and minimal analysis of the features. The main advantage of the system lies in the learning stage where it uses neural network, the text is analyzed in many stages, requires less statistical training and the ability to detect complex relationships between variables. The over-fitting problem by incorporating a feature selection method is to extract the most relevant features [Fevrie 2011]. The tweets are extracted from the Twitter API. Three kinds of features can be constructed from the tweets namely Author, Tweet and Network Features.

Tweet Features: Length of the tweet, number of words, number of unique characters, number of hash tags, number of retweets, tweet is a retweet, number of special symbols [$, !], tweet is a reply, Number of @-mentions, number of retweets, has URL, number of URLs, use of URL shortened service *Author Features*: Registration age of the user, number of statuses, number of followers, number of friends, is a verified account, length of description, length of screen name, has URL, ratio of followers to followees. *Network Features:* Degree Centrality, Eigenvector Centrality, Betweenness Centrality, Kartz Centrality.

The important features that are useful for analysis are obtained using the Chi Square feature selection method. These features are considered for further feature construction process. The resulting data is fed to the Feed Forward Neural Network to assess the credibility into one of the two classes, Credible, Not Credible using graph theory.

1. Introduction

The procedure of identifying relationship and movement between organization, people, and connected information/knowledge entities by mapping is termed as Social Network Analysis (SNA). The interpretation of network is those ponder from claiming how a group of actors socially connected between them. In this field the major study of concentration is on the data which are very comprehend. Researchers have created discrete hypothetical perspectives in this view. Some of the trademark overviews are:

- Concentrate on the connection between actors than their features;
- Sense from claiming interdependence: an atomic instead of an atomistic perspective;
- Structure influences substantive outcomes;
- Emergent impacts.

It concerns the framework principle and unpredictability hypothesis. Social networks can also be portrayed through by an individualistic technique, holding system for gathering data, measurable analysis and visual representation.

1.1. Background and History of Social Network Analysis

Social networks are networks of connections (edges) with different actors. Their size (or degree, number of nodes) is usually in graphical models that illustrate how nodes (actors) and their ties (edges, connections) form circles (clusters, cliques). Scientists can estimate density and diameter of these networks (Scott, 2000). For a given graph in Figure 1.1 extreme distance between two nodes B1 and B3 is taken as diameter represents what amount of a little reality a graph will be. "Online social networks can be represented as weighted graphs of the form: G (V, E, W) from this data, where V is the set of n nodes, E is the set of edges and W is the edge weight matrix. Let Y to be a set of m labels that can be applied to nodes of the graph" (Bhagat et al., 2011, p. 120). For example, G (V, E, W) is the form of social network which has

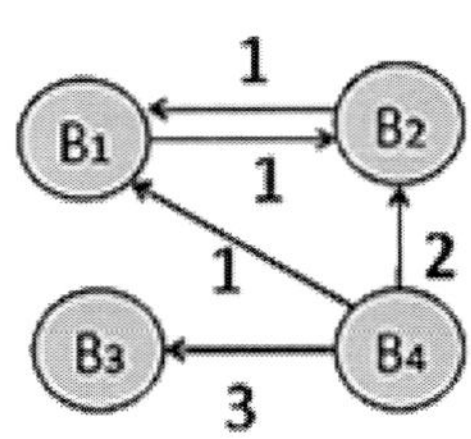

Figure 1.1. Example of social network.

- *Nodes V:* The Facebook users are represented as the set of nodes;
- *Edges* E: a relationships between two nodes could represent various possibilities such as a simple relationships (friend, relative, partner), interaction (wall post), activity (tagging in a photo), and others. Usually it is simplified to friendship;
- *Edge Weights W:* Strength of the connection is indicated by the weight (such as weak ties & strong ties);
- *Node labels Y:*

  It includes additional information such as user's interests, demographics, occupation, etc. (Bhagat et al., 2011)

  Through these indispensable components, it is shown how scientists come up with algorithms to visualize and analyze the data. The parameters are utilized in detecting central clusters and nodes in the network.
- The *three* **universal properties of social networks:**
  1) As the network increases with more nodes, its diameter condenses making the world even more trivial (McGlohon et al. 2011);
  2) They have heavy-tailed degree distribution where most nodes have few neighborhoods, and there are a few large clusters (McGlohon et al. 2011);
  3) They have strong modular or community-like structure with various interconnected clusters.

Social system dissection need its hypothetical roots in the fact that considering examples from claiming connections that associate social performing actors elucidated in the fill in for promptly sociologists. The idea from claiming "social networks" has been in usage since the early twentieth century. It is referred to the diverse sets for cooperation between parts about social frameworks and also, in addition, between nations. It needs in different academic disciplines as well as viable requisitions like countering money laundering and terrorism.

1.2. Data Used in SNA

The representation of data in SNA can be explained through simple example in Figure 1.2. Each cell has the score of some attributes and actors. For representing multiple groups third dimension of the array has to be included.

Name	*Sex*	*Age*	*In-Degree*
Bob	Male	32	2
Carol	Female	27	1
Ted	Male	29	1
Alice	Female	28	3

Figure 1.2. Rectangular data array.

By correlating and comparing columns the similarity between the variables are found in their distribution across the actors "Network" data which is represented in a square array of measurements (Santana et al. 2011)

Row = subject or case or observation

Column = same set of subjects, case, or observations cell connection between actors.

Who reports liking whom?				
	Choice:			
Chooser:	Bob	Carol	Ted	Alice
Bob	---	0	1	1
Carol	1	---	0	1
Ted	0	1	---	1
Alice	1	0	0	---

Figure 1.3. A square array of network data.

From the Figure 1.3 questions that can be answered are whom they choose, and second one is who in terms of being chosen by others. Former question is answered by comparing the rows in the array, so that which actor is similar to which other actor is known and the later one is solved by comparing the similarity between the columns. Analyzing the data aides to viewing the actors who have comparable positions in the system. This will be those urgent stresses about system analysis: how actors are placed or "inserted" in the general organization.

The information structure can be analyzed by an alternate approach. The rise in numbers of 1's and 0's in the matrix infers that there is a moderate "density". The cells above and below the diagonal can be compared to see if there is reciprocity in choices This may be the second major stress of organizational analysis: perceiving how the entirety design of singular decisions provides for the climb will more comprehensive examples.

1.3. Network Theory

Identification of symmetric or asymmetric relation between objects can be examined by a graph of that particular network. It may be a piece of graph theory in which hubs or edges are the attributes (e.g., User-id). It is being utilized in numerous areas including computer science, electrical engineering, biology, economics, finance, operations research, climatology, sociology, logistical networks, web, Internet, gene administrative networks, etc. (Resnick et al. 1994).

1.3.1. Data Collection

Collection of data is one of the important tasks in analyzing the network. Depending on the data collected the result will be interpreted. Data collection is defined as a collection of data on target variable in a systematic way.

A. Observation

It is based on the researchers' own examination of data, preferably in a natural setting which is a systematic one.

B. Using Documents

It is the collection of data from already existing data documents. For example sales, purchase, customer information, etc. It is one of the easiest ways of collecting data.

2. GRAPH THEORETIC NETWORK

2.1. Introduction

The relations between the set of objects in graphs are identified using a proper defined structure. It is termed as Graph theory. It can be vertices, nodes, or points which are connected by edges, arcs, or lines which are shown in Figure 2.1.

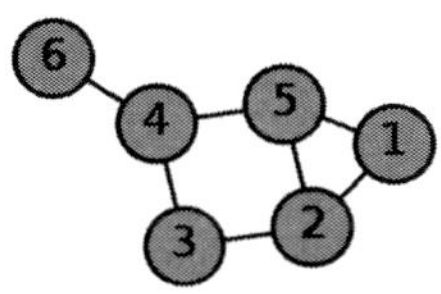

Figure 2.1. Graph.

2.2. Organizing and Analyzing the Data

Organizing the Data

Organizing the data, designing, analysis plan and frequency of specific behavior are the important point to be noted. Qualitative data is non-numerical information, such as responses gathered through unstructured interviews, observations, focus groups, or open-ended survey questions (Mojtaba Saleh et al. 2014).

Analyzing the Data

Statistical analysis of quantitative information can be tedious. Descriptive analysis can reduce raw data down to an understandable level like -

- Frequency distributions – charts or tables;
- Central tendency - mode (number or category that appears most frequently), median (number in the exact middle of the data set), and mean (arithmetic average of your numbers);
- Variability – range (difference between the highest and lowest scores) and standard deviation (a more complicated calculation based on a comparison of each score on the average).

Using probability theory, the significant result will have high probability, showing that it was not due to chance, inferring that a relationship between two variables is strong and reliable. Factors influencing the likelihood of significance include the strength of the relationship, the amount of variability in the data, and the number of people in the sample (Leskovec et al. 2010). When the data available are for a small number of people, investigations deal with the clinical significance. It is the comparison of results to a pre-established standard that has been determined to be meaningful (average functioning of a non-problematic peer group, cultural norms, or goals set by staff or participants).

2.3. Matrices and Relational Data

A set of elements which is arranged in a rectangular fashion in rows and columns is called matrix. An "m by n" matrix has m rows and n columns. Special case of matrix exists in which if it has single row then it is called "row vector and if it has single column it is called a column vector". Figures 2.2 and 2.3 shows the example of 2 X 4 and 4 X 4.

1,1	1,2	1,3	1,4
2,1	2,2	2,3	2,4

Figure 2.2. Example: "2 X 4" matrix.

1,1	1,2
2,1	2,2
3,1	3,2
4,1	4,2

Figure 2.3. Example: "4 X 4" matrix.

Each cell in the matrix has their address. For example Element (4, 1) is the entry in the fourth row and first column as shown in Figure 2.4.

	A	B	C	D
A	---	1	0	0
B	1	---	1	0
C	1	1	---	1
D	0	0	1	---

Figure 2.4. 4 X 4 with additional row and column labels.

Mostly in a social network, square matrix is used in which it has an equal number of rows and columns. In some cases, rectangular matrices also 3-dimensional are also used.

The Adjacency Matrix

Adjacency matrix is used to know about, who is adjacent or next to whom, and the initial point of the network. It is one of the most common matrix is binary. It can be "symmetric" or "asymmetric". The directed graph of friendship choices among Bob, Carol, Ted, and Alice is shown in Figure 2.5 is a simple example for adjacency matrix representation.

	Bob	Carol	Ted	Alice
Bob	---	1	1	0
Carol	0	---	1	0
Ted	1	1	---	1
Alice	0	0	1	---

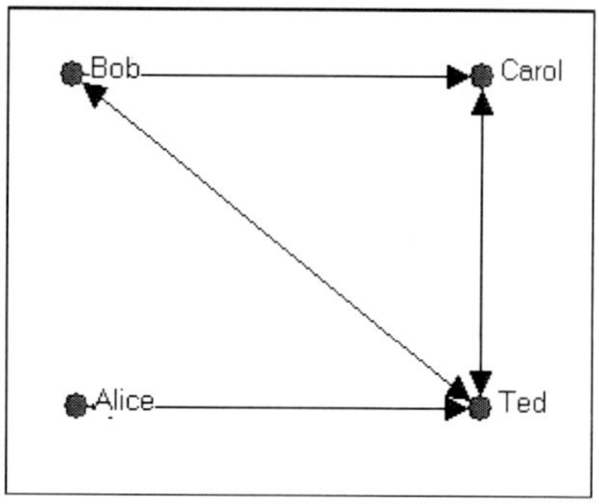

Figure 2.5. Adjacency Matrix.

The ties are measured at the nominal level (that is, the data are binary choice data). The adjacency matrix is a square, 1-mode actor-by-actor matrix representation as shown in Figure 2.6 and Table 1.1.

Table 1.1 Actor by actor matrix

	Andy	Bill	Carol	Dan	Elena	Frank	Garth
Andy		1	0	1	0	0	1
Bill	1		1	0	1	0	0
Carol	1	1		1	1	0	0
Dan	1	1	1		0	0	0
Elena	0	0	0	0		1	0
Frank	0	0	0	0	1		0
Garth	1	1	0	0	0	0	

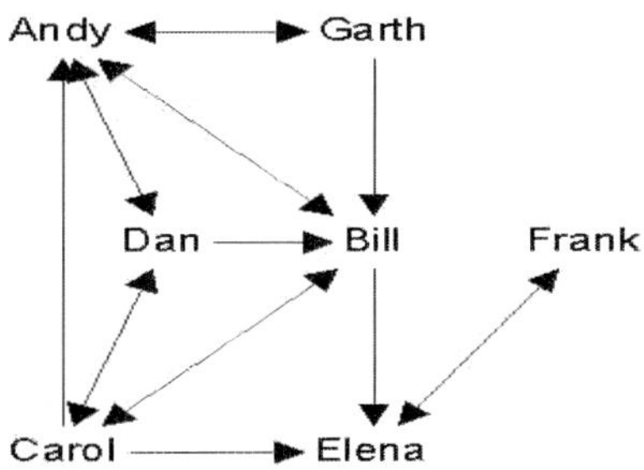

Figure 2.6. Actor by actor matrix representation.

2.4. Matrix Conventions

If something is represented in the graph, then it can be represented in the form of a matrix, e.g., a valued graph contains zeros and ones in the matrix. When a matrix is coined as K then its cells are denoted as K_{ij}. Generally data is recorded so that the "row" conforms to "column". For example, *i* and *j* be the row and column and relation between them is "getting advice from" then the value of $K_{ij} = 1$ which interprets that *i* is getting advice from *j*. If any changes have to be made in rows and columns, i.e., to interchange the rows and columns, then transpose matrix has to be used which is said to be K'.

Matrix Algebra

In matrix algebra, rows define the people and columns define their properties. Different class of entities in a matrix is termed as mode. The representation of the matrix will always be in capital letters like small letter representation is the specific cell number. In general a matrix K with row m and column n is represented as K_{mn}.

Different mathematical operations can be performed on the matrix like addition and subtraction. For this, the matrix should be in the same size. Matrix multiplication is slightly different from the other operations since multiplication in matrix is not commutative. (AB = BA - not always true). The number of columns in the first matrix must equal the number of rows in the second. The product that results from a matrix multiplication has as many rows as the first matrix and as many columns as the second.

For example:

X =

1	2
2	4
1	0

Y =

1	1	2	2
2	2	4	4

Z = XY =

5	5	10	10
11	11	22	22
1	1	2	2

Y*X is not possible because they are not conformable.

3. Terminology for Network Analysis

3.1 The Language of Network Analysis

Social networks speak of cooperation around social substances similar to companionships between individuals, correspondence in a group, transactions among enterprises. Distinguishing those principle actors, finding durable assemblies and aid samples for is used for analyzing social networks. High-level capacities make it easy to recognize communities, find durable groups, and visualize the effects. A full suited social organize measures make it workable to investigate networks, rank actors from their centralities, or give proposals dependent upon similar artist (Lipczak et al. 2009).

3.2. More Than Joining Up the Lines

The majority of tools of social system investigation manage a solitary sort of association around actors: friendship, kinship, monetary exchange, warfare, and so forth. However, social relations around actors are more unpredictable viz-a-viz gatherings of persons; it might be a chance to be enthusiastic connections, return relations, connection ties (Lin et al. 2009) Furthermore, different associations may act for the same duration of time. Two general methodologies followed are combination and reduction "Reduction" looks to consolidate data around numerous relations around the same set about on-screen characters under a single connection that indexes the quantity of binds. "Combination" makes an absolute list of the multiplex relations, yet all the endeavors will representable the quality of binds.

3.3. The Flow of Information and Resources

In a disengaged environment, information has to be sent through hub encounters in order to send the data to a destination. Strategies should be connected to the setting for directing in DDTMs. Graphs might make used to mean those social structure about social networks done in a characteristic way – hubs, likewise, is a vertebrate fossil science of a chart

Centrality over graph hypothesis and analyzing the network will be a quantification of the relative fact that a vertebrate fossil science inside the chart (for example, how essentially an individual will be in a social network). A central hub needs a stronger proficiency of interfacing with other organizational parts. Freeman's degree, closeness, betweenness measures would use to measure centrality.

3.4. Density of Connections

In a network, the things that associated would be called as "nodes". A hub might be a person, a computer, or actually a portion hyperlinked content.

The associations between hubs may be "connections" and frequently called "edges". "Network density" depicts the part for the "potential connections" on a system that are "actual connections". A "potential connection" is an association that might conceivably exists the middle of two "nodes" (Hsu et al. 2006). An "actual connection" is one that really exists.

In Figure 3.1 the network A and B demonstrate the case that the number of the original connections between the nodes will be as same as the number of potential connections.

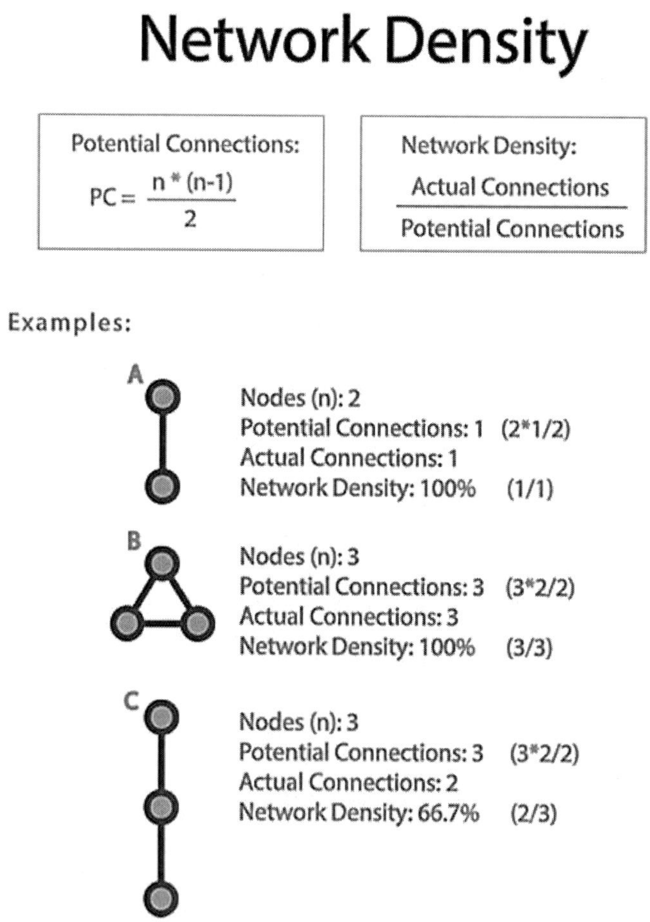

Figure 3.1 Calculating Network Density.

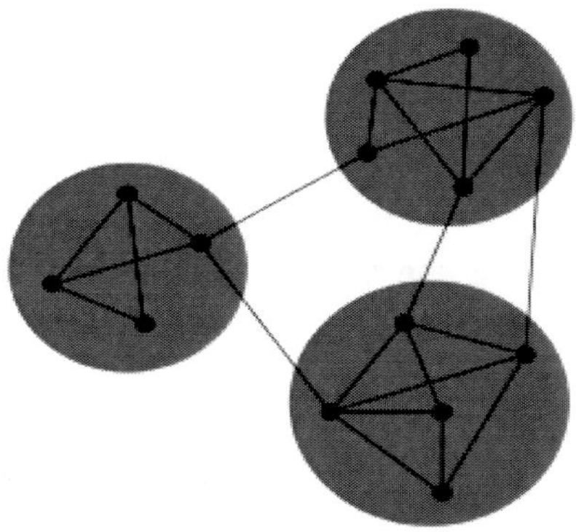

Figure 3.2 Community structure.

3.5. Community Structure and Density

In a network if the nodes are connected internally in a densely manner it is said to possess the "community structure" as shown in Figure 3.2. Even though it is overlapped, it has the principle that if the nodes are in the same community, then it is highly connected to each other and less likely to be connected if they do not share communities. Community search is to find out to which community cervix vector it belongs to. Example of Social networks includes groups based on common university, interests, occupation, etc.

4. CENTRALITY

4.1. Local and Global Centrality

This concept of identifying the important vertices in a graph based on the ranking, which in turn produced by the values is called centrality (Wang et al. 2010).

An essential issue is the one 'at the Centre' of a number of associations. The simplest technique with recognizing is toward those degrees of the different focuses in the chart. A side of the point is neutral, assuming that it needs a secondary degree (Ulrik Brandes et al. 2001). Degree-based measures in neighborhood centrality might additionally make points for focuses in guided graphs: if done these particular circumstances every perspective will bring two measures of its local centrality, person comparing should note it's in degree and the difference to its out degree. Figure 4.1 shows the local and global centrality (El-Hefnawy et al. 2013) those important 'neighborhood' is widened will incorporate those greater amount inaccessible associations of the focuses.

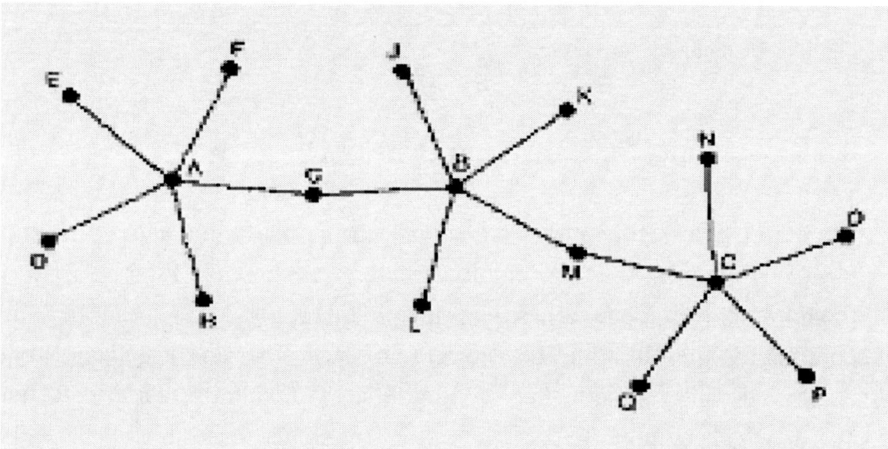

		A, C	B	G, M	J, K, L	All other points
Local centrality	Absolute	5	6	2	1	1
	Relative	0.33	0.33	0.13	0.07	0.07
Global centrality		43	33	37	48	57

Figure 4.1. Local and global centrality.

4.2. Betweenness Centrality

Betweenness centrality refers to the number of times a node acts as a bridge along the shortest path between two other nodes. It was introduced as a measure for quantifying the control of a human on the communication between other humans in a social network.

For the connected network, the betweenness of a vertex v in a graph G: = (V, E) with V vertices is computed as follows (Silva et al. 2001): For each pair of vertices (s, t), compute the shortest paths between them. For each pair of vertices (s, t), determine the fraction of shortest paths that pass through the vertex shown in Figure 4.2. Sum this fraction over all pairs of vertices (s, t). More compactly the betweenness can be represented in Figure 4.3 and computed as

$$Betweeness\left(V\right)=\sum s6=v6=6\in V\sigma st(v)\sigma st \qquad (1)$$

where σst is the total number of shortest paths from node ‘s’ to node ‘t’ and σst(v) is the number of those paths that pass through v.

The betweeness may be normalized by dividing through the number of pairs of vertices not including v, which for directed graphs is (n – 1)

(n – 2) and for undirected graphs is (n – 1) (n – 2) / 2.

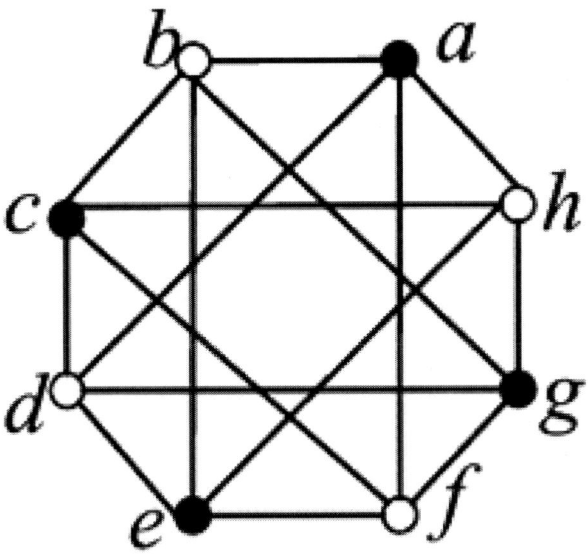

Figure 4.2. Connected network

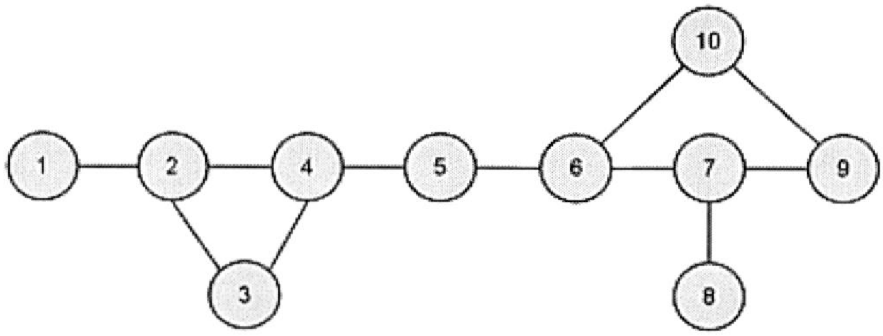

Figure 4.3. Example of a network for the illustration of centrality measures.

4.3. Centrality Boosts Centrality

Degree Centrality (DC) σ_D represents the simplest Centrality Measure (CM) and determines the number of direct contacts as an indicator of the quality of a network member's interconnectedness. Using the adjacency matrix $A = (a_{ij})$ it can be defined as

$$\sigma D(x) = \sum I = \ln(a_i(x)\sigma D(x) \quad (2)$$

As a consequence, the centrality score $\sigma_D(x)$ for a node x is higher, the more contacts a node x has. In the network it follows that $\sigma_D(1) = 1$ since actor 1 has only one direct relationship with actor 2. In contrast, actor 4 has the centrality score $\sigma_D(4) = 3$.

Table 4.1 shows the values of DC for all members of the example network. In addition, the actors' ranking (in short "rank") is stated, i.e., their order in descending value of DC. The actors 2, 4, 6, and 7 take rank 1 and thus are the best networked members when applying this CM (Echegaray et al., 2009).

Table 4.1. Degree centrality

Actor x	1	2	3	4	5	6	7	8	9	10
σD(x)	1	3	2	3	2	3	3	1	2	2
Rank	9	1	5	1	5	1	1	9	5	5

The major disadvantage of DC is that indirect contacts are not considered at all. Therefore, a reduction of the distance from one actor *x* to another actor *y* resulting from an additional relationship in most cases does not increase the value of the CM. The intensification of a connection of shortest length between *x* and *y* does also not increase the value of this CM, since DC only considers direct contacts. However, in an undirected, unweighted graph a direct connection between the actors *x* and *y* can exist only once. Through a new relationship, both actors involved win one additional direct contact. So the DC of both members equally increases by 1 and the ranking of the actors thus always remains the same.

4.4. Centralization and Graph Centre

The formal properties claiming centrality were at first investigated by Bavelas (1950). Similarly, as an aftereffect for this burgeoning of formal measures of centrality, there will be respectable perplexity in the territory. A point will be comprehensively central, on the different hand, as it needs a position of vital importance in the general structure of the organization. Nearby centrality will be concerned with the relative unmistakable quality of a point of convergence (Domingos et al. 2001). Previously, its neighborhood, same time worldwide centrality worries unmistakable quality inside the entire organize.

5. Subgraphs, Groups and Social Divisions

5.1. Identifying Subgraph

On a social graph, vertices speak to social actors, same time edges speak to connections or cooperation between on-screen characters. An essential operation is on recognize gatherings for social on-screen characters that would exceedingly be associated with every other, spoken to by a durable sub graph, alongside which structural examples involving social actors.

Even ordinary clients might recognize the thing that going on in their neighborhood.

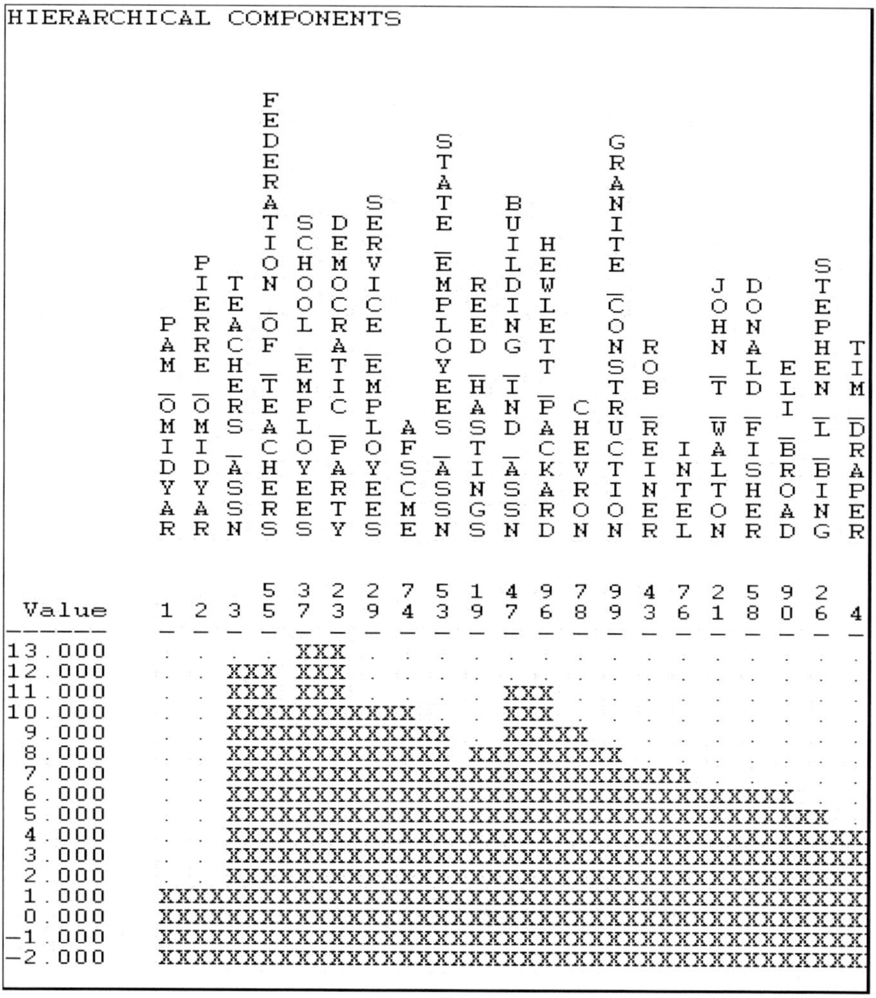

Figure 5.1. Hierarchical components.

5.2. The Components of a Network

Parts of a graph would sub-graph what need to be joined within, however disengaged between sub-graphs. The guided graphs (in contrast on straightforward graphs), can be characterize two separate components. A solid part obliges that there is a guided way between nodes with the end goal to make in the same component. A feeble part will be a situated from hubs connected without the direction of ties. The quality of the connection between two performing actors may be characterized toward those numbers for times that they contributed to the same side of an issue. Figure 5.1 shows some of the hierarchical component values.

5.3. Cycles and Circuits

A simple cycle may be characterized as shut stroll for no reiterations of vertices, edges can be allowed, other than those beginning and ending vertex alternately. Concerning illustration the situated about edges in such a stroll. A circuit can make a shut stroll permitting reiterations from claiming vertices anyway not edges. Figure 5.2 shows how circuits help in better community finding.

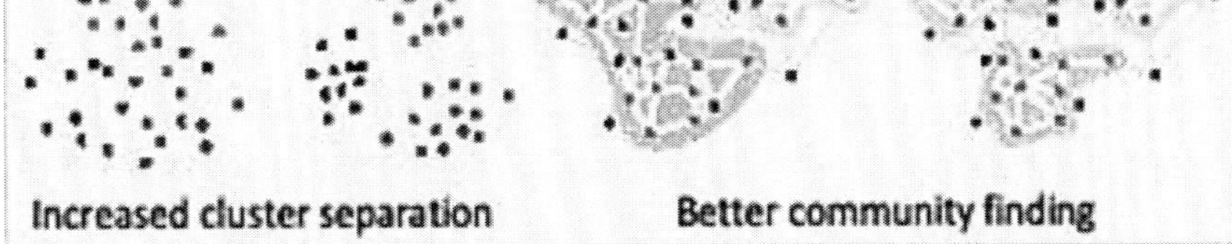

Figure 5.2. Community finding.

5.4. Cliques within Components

It may be a subset of vertices for an undirected graph in which each two different vertices in the clique would adjacent; that is, its induced sub graph is complete.

A clique, C, Previously, an undirected graph G = (V, E) is a subset of the vertices, C and V, such-and-such each two dissimilar vertices are contiguous. This may be equal of the condition that the induced sub graph of G induced by C is a complete chart (Fevrier et al. 2011). The expression coterie might additionally allude of the sub graph specifically.

A maximal clique is an inner circle that can't a chance to be developed that is, an inner circle which doesn't exist only inside the vertebrate fossil science situated of a bigger coterie. Some writers define cliques as it were, that obliges them with a chance to be maximal, and also the utilization of different wording for complete sub graphs need not be maximal.

A maximum clique of a graph, G, will be a clique, such-and-such there may be no clique with All the more vertices. Moreover, the clique number (G) is those vertices number alongside a most extreme clique in G. The intersection number of G is that little amount for cliques that together, spread the sum edges of G.

The smallest number from cliques of G whose union blankets the set about vertices V of the chart is said to be the clique blanket number of G. A maximum clique transversal of a chart will be a subset of vertices for that property that every greatest coterie of the chart holds in any event person vertebrate fossil science in the subset (J Beel, et al. 2013). The inverse of a coterie is an independent set, in the feeling that each clique corresponds on an autonomous set in the complement chart. The clique cover problem worries discovering similarly as couple cliques similarly as workable that incorporate each vertebrate fossil science in the chart.

A related particular idea is a biclique, a complete bipartite sub graph. The bipartite dimension of a chart will be the least amount for bicliques necessary to spread every last one of edges of the chart.

6. STRUCTURAL LOCATIONS, CLASSES AND POSITIONS

6.1. The Structural Equivalence of Points

In a network any two vertices share more number of same neighbors, and then those vertices are called structurally equivalent, which is said to be the strongest form of similarity (J Beel, et al. 2014). In Figure 6.1 E, F, H, I shares the same number of neighbors.

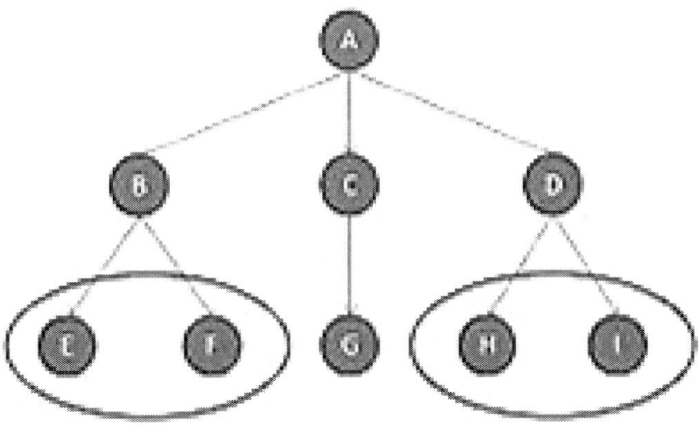

Figure 6.1. Example of Structural equivalence.

6.2. Clusters and Similarity

Similarity in analyzing the network happens when two hubs fall into those same comparability classes. There are three key methodologies for constructing measures for system similarity: regular equivalence, automorphic equivalence, structural equivalence (Wasserman et al. 2004). There is a progressive structure of the three proportionality concepts: whatever is of structural equivalences is likewise automorphic and regular equivalent. Any automorphic equivalences are also regular equivalences (Mojtaba Salehi et al. 2014). Not the greater part general equivalences would fundamentally automorphic and not all automorphic equivalences need to be fundamentally structural.

6.3. Divide and Conquer

Divide and conquer is one of the famous algorithm in which a problem is divided into sub problems and each sub problem is conquered finally combined to produce the solution for a problem.

7. Social Changes and Developments

7.1. Small World Networks

The small-world network is a special type of graph shown in Figure 7.1 in which most hubs would not have neighbours from claiming hub, yet the neighbours for any hub are disposed to make neighbors about one another (Sarwar et al. 2001). Any hub can be reached from any other hub with the least number of leaps or jumps. The node distance between any two randomly selected nodes grows proportionally to the logarithm of the number of nodes in the network. Examples of small world networks are Voters network, Route maps, and mobile phone call graphs.

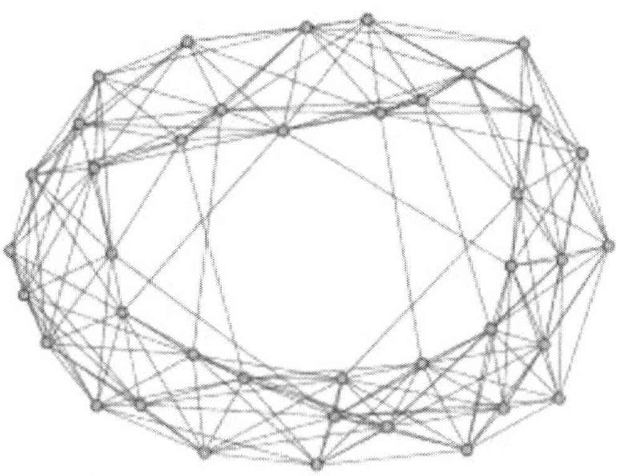

Figure 7.1. Example of small world network. {\displaystyle L\propto \log N}.

7.2. Modeling Social Change

The concept of social networks and outcome of the analysis, representing it visually is imperative. The data in the network can be expolarated through the nodes which are displayed and also through various properties like colors, different layouts and various other properties of the node (Conrad Lee et al. 2010)

7.3. Visualizations

Social system Investigation will be a methodology in the social sciences utilizing graph-theoretic ideas, should describe, furthermore illustrating social structure. The key line toolkit and vision software can be incorporated in the Investigation and visualization of social networks.

The following are the some of the basic terms in SNA visualization

- *Modularity* - measure of the structure of networks or graphs. It was designed to measure the strength of the division of a network into modules;
- *Average Distance* - The Average of distance between all pairs of nodes;
- *Connected Components* - a connected component (or just component) of an undirected graph is a sub graph in which any two vertices are connected to each other by paths, and which is connected to no additional vertices in the super graph;
- *Average Degree* - Average number of links per node;
- *Distance* - The number of edges along the shortest path connecting them;
- *Network Diameter* - The maximum distance between any pair of nodes in the graph;
- *Average Weighted Degree* - Average of sum of weights of the edges of nodes.

CONCLUSION

Social network data are the most important data which provide the information about the emergency events like flood, earthquake, tsunami and other natural calamity. Based on the social network data, volunteers can rescue the people and provide the necessary assistance. Hence, procession of social network data are vital functions with respect to different parameters like author, network and the type of social networks.

Graph theory plays an important role to process such social network data. Especially Degree Centrality (DC), Betweenness Centrality are the source of the analysing the social network data about the origin, destination and intermediate communications (Data Transfer). The Graph theory can be applied to find the importance of data floated on the social network and also help to identify the correct way of rescuing activity. In addition, a neural network can be used as a credibility measure of the social network data.

REFERENCES

Beel, J., Langer, S., Genzmehr, M., Gipp, B. And Nürnberger, A., 2013. A Comparative Analysis of Offline and Online Evaluations and Discussion of Research Paper Recommender System Evaluation, In *Proceedings of the Workshop on Reproducibility and Replication in Recommender Systems Evaluation (RepSys) at the ACM Recommender System Conference (RecSys), ACM International Conference Proceedings Series (ICPS)*, pp. 7–14.

Beel, J., Langer, S., Genzmehr, M., Gipp, B., Breitinger, C. And Nurnberger, A., 2013. Research Paper Recommender System Evaluation: A Quantitative Literature Survey, In *Proceedings of the Workshop on Reproducibility and Replication in Recommender Systems Evaluation (RepSys) at the ACM Recommender System Conference (RecSys), ACM International Conference Proceedings Series* (ICPS), ACM, pp. 15–22.

Beel, J., Langer, S., Gipp, B. and Nurnberger, A., 2014. The Architecture and Datasets of Docear's Research Paper Recommender System. In *Proceedings of the 3rd International Workshop on Mining Scientific Publications (WOSP 2014) at the ACM/IEEE Joint Conference on Digital Libraries* (JCDL 2014).

Conrad Lee, Fergal Reid, Aaron McDaid, Neil Hurley, Detecting highly overlapping community structure by greedy clique expansion, from *Research Repository,* UCD, Clique Research Cluster, University College Dublin, Ireland, July 2010. http://hdl.handle.net/10197/2516.

Domingos and M. Richardson, Mining the Network Value of Customers, *Proceedings of the Seventh International Conference on Knowledge Discovery and Data Mining*, San Francisco, CA, 2001. ACM Press, pp. 57-66.

Echegaray, J. Morales, and W. Luo, Optimizing social life using online friend networks, in *System of Systems Engineering, 2009. SoSE 2009, IEEE International Conference on*, May 2009, pp. 1 –5.

El-Hefnawy N. A., Solving Bi-level Problems Using Modified Particle Swarm Optimization Algorithm, *International Journal of Artificial Intelligence,* vol.12, no.2,pp. 1781-1794,,2013.

Fevrier Valdez, Patricia Melin, Oscar Castillo, An improved evolutionary method with fuzzy logic for combining Particle Swarm Optimization and Genetic Algorithms, *Applied soft computing,* vol.11, no.2,pp. 2625-2632, 2011.

Hsu, A. King, M. Paradesi, T. Pydimarri, and T. Weninger., Collaborative and structural recommendation of friends using weblog-based social network analysis", *Proc. of AAAI Spring Symposium Series,* 2006.

Lin, L. Wang, and S. Guo, Recommendations on social network sites: From link mining perspective, in *Management and Service Science, 2009. MASS '09. International Conference on*, September 2009, pp.1 –4.

Lipczak and E. Milios, Agglomerative genetic algorithm for clustering in social networks, in *Proceedings of the 11th Annual conference on Genetic and evolutionary computation*, GECCO '09. New York, NY, USA: ACM, 2009, pp. 1243–1250.

Leskovec, K. J. Lang, and M. W. Mahoney, Empirical comparison of algorithms for network community detection, *Proceedings of the 19th international conference on World wide web WWW 10*, vol. 30, p. 631, 2010.

Mojtaba Salehi, Isa Nakhai Kamalabadi and Mohammed Bagher Ghazavi Ghoushci, Personalized Recommendation Of Learning Material Using Sequential Pattern Mining And Attribute Based Collaborative Filtering, *Education and Information Technologies,* Springer, December 2014, Volume 19, Issue 4, pp. 713-735

Resnick, N. Iacovou, M. Sushak, P. Bergstrom, and J. Riedl, Grouplens: An open architecture for collaborative filtering of net news, *Proceedings of CSCW 1994,* ACM SIG Computer Supported Cooperative Work, 1994.

Santana, C. Bielza, and P. Larraaga, Optimizing brain network topologies using multi objective evolutionary computation, *Neuroinfor; matics, vol. 9,* pp. 3–19, 2011, 10.1007/s12021F010F9085F7.

Sarwar, G. Karypis, J. A. Konstan, and J. Riedl, Item-based collaborative filtering recommendation algorithms, *Proceedings of the 10th International World Wide Web Conference* (WWW10), Hong Kong, May 2001.

Silva, I. F. R. Tsang, G. D. C. Cavalcanti, and I. F. J. Tsang, A graph based friend recommendation system using genetic algorithm, in *Evolutionary Computation (CEC), 2010 IEEE Congress on*, July 2010, pp. 1 –7.

Ulrik Brandes, A faster algorithm for betweenness centrality - *Journal of Mathematical Sociology,* 2001, Taylor & Francis Vol. 25(2), pp. 163-177.

Wang, Y. Tan, and M. Zhang, Graph based recommendation on social networks, in *Web Conference (APWEB), 2010 12th International Asia; Pacific*, April 2010, pp. 116 –122.

Wasserman, K. Faust, *Social Network Analysis: Methods and Applications,* Cambridge University Press, (1994).

BIOGRAPHICAL SKETCH

V. Pattabiraman, PhD (author)

Affiliation: School of Computing Science and Engineering, Vellore Institute of Technology University, Chennai, India.

Education: PhD

Business Address: Associate Professor, School of Computing Science and Engineering, Vellore Institute of Technology University - Chennai Campus, Chennai-600127, India.

Research and Professional Experience: 18 Years

Professional Appointments:

1. Professional Member of ACM, ACM Membership No.: 1140600, since 2011.
2. Professional Member of Academy & Industry Research Collaboration Center (AIRCC), membership id: 128785, since 2007.
3. Professional Member of IACSIT, since 2009.

Honors:

Received a Springer Best Paper Award "Self-Repairing Functional Unit Design in an Embedded Out-of Order Processor Core" -Paper ID: CS209 during Springer International Conference on Computational Sciences, Advanced Database and Computing (CSADC2017) held on11-12, October, 2017 at Phuket, Thailand.

Received an award for the Academic Excellence and Research from VIT University, Chennai, during the academic year 2013-14.

Received a Research award for the Research paper Publication in Peer reviewed Journals, from VIT University, Chennai, during the academic year 2015-16.

Publications from the Last 3 Years:

1. Suresh, K. and Pattabiraman V. (2015), Investigation on ARM Algorithm with Incremental Mining and Proposing a New Algorithm for Streaming Dataset, *Australian Journal of Basic and Applied Sciences*, 9(6) Special 2015, Pages: 72-76.
2. Suresh, K. and Pattabiraman, V. (2015), An Improved Utility Itemsets Mining with respect to Positive and Negative Values Using Mathematical Model, *International Journal of Pure and Applied Mathematics*, Volume 101 No. 5 2015, 763-772.
3. Muralidhar, A. and Pattabiraman V. (2015), An Efficient Association Rule Based Clustering of XML Documents, *Procedia Computer Science* 50 (2015) 401 – 407. Published by Elsevier.
4. R. Maheswari, Pattabiraman V. (2015), Performance Analysis of Reconfigurable Open Soft Core Processor-OR1200 with Global Hazard Controller, *International Journal of Pure and Applied Mathematics*, Volume 101 No. 5 2015, 849-858.
5. Graceline Jasmine S. and Pattabiraman V. (2015), Hyperspectral Image Analysis using End Member Extraction Algorithm, *International Journal of Pure and Applied Mathematics*, Volume 101 No. 5 2015, 809-829.
6. Harini S., Pattabiraman V. (2015), Hybrid Multi-Core ASIC Architecture with FPGA for Self-Repairing of Hard Faults, *International Journal of Pure and Applied Mathematics*, Volume 101 No. 5 2015, 905-913.
7. Abirami, M., Pattabiraman, V. (2016)., Data mining approach for intelligent customer behavior analysis for a retail store, *Smart Innovation, Systems and Technologies*, Vol. 49 pp. 283-291, March, 2016. © Springer International Publishing Switzerland 2016.
8. Muralidhar A. and Pattabiraman V. (2016), A Novel approach to extract greater profitable itemsets, *International Journal of Pure and Applied Mathematics*, Volume 109 No. 8, 2016, 83 - 90 ISSN: 1311-8080 (printed version); ISSN: 1314-3395.
9. Maheswari R. and Pattabiraman V. (2016). Petri Nets Based System Modeling for Performance Analysis of Reconfigurable Soft-core

Processor Using FPGA, *International Journal of Technology*, Volume: 7, Issue: 6, pp: 1054-1067.

10. Mohammed R. K. and Pattabiraman V. (2016), Curing the Curse of Dimensionality with Multi-fold Deep Denoising Autoencoder, , *International Journal of Control Theory and Applications*, Volume 9, Number 51, pp. 113-122, Dec 2016, International Science Press, ISSN : 0974-5572.
11. Sivagama Sundari R. and Pattabiraman V. (2016), A Complete Investigation to Demystify the Math Behind PCA using the MNIST Dataset, *International Journal of Control Theory and Applications*, Volume 9, Number 51, pp. 249-260, Dec 2016, International Science Press, ISSN : 0974-5572.
12. Graceline Jasmine S. and Pattabiraman V. (2016), Performance Evaluation of Dimensionality Reduction in Hyperspectral Imaging, *International Journal of Control Theory and Applications*, Volume 9, Number 52, pp. 168-178, Dec 2016, International Science Press, ISSN : 0974-5572.
13. Harini Sriraman and Pattabiraman Venkatasubbu (2017), On the field design bug tolerance on a multi-core processor using FPGA, *Int. J. High Performance Computing and Networking*, Vol. 10, Nos. 1/2, pp. 34-42, 2017. (InderScience)
14. Maheswari R. and Pattabiraman V. (2017), Improved reconfigurable hyper-pipeline soft-core processor on FPGA for SIMD, *Int. J. High Performance Computing and Networking*, Vol. 10, No. 3, pp. 207-217, 2017(InderScience)
15. Maheswari R., Pattabiraman V, Sharmila P. (2017), Reconfigurable FPGA Based Soft-core Processor for SIMD Applications, *Asian Journal of Pharmaceutical and Clinical Research*, Special issue, April, pp: 180-186.
16. Reshma Remesh, Pattabiraman V. (2017), A Survey on the Cures for the Curse of Dimensionality in Big Data, *Asian Journal of Pharmaceutical and Clinical Research*, Special issue in Advances in Smart Computing and Bioinformatics, April-2017, pp: 354-360.

17. Harini Sriraman and Pattabiraman V. (2017), Self-Repairing Functional Unit Design in an Embedded Out-of Order Processor Core, *The International Conference on Computational Sciences, Advanced Database and Computing (CSADC2017)*, Springer book series on Advances in Intelligent Systems and Computing (ISSN: 2194-5357).
18. Mehfooza, M. and Pattabiraman V. (2017). An assessment on domain ontology-based information extraction techniques, *Int. J. Services Technology and Management*, Vol. 23, No. 4, pp.299–312.
19. Mehfooza M. and Pattabiraman V. (2018), Analysis of Supervised Machine Learning Techniques on the Application of the Indian Coastal Dataset, *International Journal of Pure and Applied Mathematics*, Volume 118, No. 18, 2018, pp.1913-1923.

R. Parvathi, PhD (co-author)

Affiliation: School of Computing Science and Engineering, Vellore Institute of Technology University, Chennai, India.

Education: PhD

Business Address: Associate Professor, School of Computing Science and Engineering, Vellore Institute of Technology University - Chennai Campus, Chennai-600127, India.

Research and Professional Experience: 18 Years

Professional Appointments:

1. Professional Member of International Association of Computer Science and Information Technology
2. Professional Member of International Association of Engineers

Honors:

Received an award for the Academic Excellence and Research from VIT University, Chennai, during the academic year 2013-14.

Received a Research award for the Research paper Publication in Peer reviewed Journals, from VIT University, Chennai, during the academic year 2015-16.

Paper Publications from the Last 3 Years:

1. Anto Rosaline, Parvathi R., Performance Analysis of Various Text Classification Algorithms, *International Journal of Pure and Applied Mathematics*, Volume: 101, Issue: 5, Aug, 2015, Page Nos: 625-634.
2. Rekha J. H., Parvathi R., Survey on Software Project Risks and Big Data Analytics, *Procedia Computer Science* 50 (2015) 295 – 300.
3. Anto Rosaline, Parvathi. R. Mood Based Recommendation System in Social Media Using PSV Algorithm, *Smart Innovation, Systems and Technologies*, Volume: 49, Issue: 0, Feb, 2016, Page Nos:345-352.
4. Bangara Kamatchi, Parvathi R. Call Log Analysis Using Different Clustering Approach, *International Journal of Engineering Science and Computing*, Volume: 6, Issue: 6, June, 2016, Page Nos:6952 – 6954.
5. R.Vinodha and Parvathi R. Location Based Travelling Recommendation System Using Hypertext Induced Topic Search (HITS), *International Journal of Pure and Applied Mathematics*, Volume 109 No. 9 2016, 248– 256.
6. Vignesh U, Parvathi R. Clustering on Structured Proteins with Filtering Instances on Bioweka, *Journal of Engineering Science and Technology*,Vol. 12, No. 3 (2017) 820 – 833.
7. Joshika, Parvathi R. A Survey on Business Facilities Analysis, *Asian Journal of Pharamacutical and Clinical Research*, 04, 2017, Page Nos:15-19.
8. Merin Susanna Verghess, Parvathi R. A Survey on Usage Analytics for Imaging Diagnostic Equipments, *Asian Journal of Pharamacutical and Clinical Research*, 04, 2017, Page Nos:457-460.

9. Bangara Kamatchi, Parvathi R. Feature Selection and Yield Prediction of Rice Crop using Ann. *Journal of Advanced Research in Dynamical and Control Systems*, Vol. 9, No 12.2017, pp.546-553.

Book Chapters Published:

1. Parvathi R. & Bangara Kamatchi, A Review on Spatial Big Data Analytics and Visualization published in *Modern Technologies for Big Data Classification and Clustering.* ISSN/ISBN Nos: 9781522528050, Page Nos: 179-197, IGI Global 2017.
2. Parvathi R. & Vignesh U., Biological Big Data Analysis and Visualization published in *Modern Technologies for Big Data Classification and Clustering.* ISSN/ISBN Nos: 9781522528050 Page Nos: 244-259, IGI Global, 2017.

In: Network Theory and Analysis
Editor: Amber Jorgensen
ISBN: 978-1-53613-460-5

Chapter 4

AN INTRODUCTION TO IPV4 ADDRESSING SCHEMES, MODES AND CLASSIFICATIONS

***A. Dhanapal*[1,*] *and V. Vijayakumar*[2]**
[1]School of Computing Science and Engineering,
VIT University, Chennai, India
[2]School of Computing Science and Engineering,
VIT University, Chennai Campus, India

ABSTRACT

This chapter provides an insight into an IP addressing, and different IP addressing schemes available today. Upon completing this chapter, the reader shall understand the details on IPv4 addressing, the basic concepts behind the IP addressing schemes such as Host ID, Network ID, Unicast addressing, Multicast addressing, Broadcast addressing, classifications of the IPv4 addressing schemes, Subnetwork, Subnet mask and default subnet mask. The details on the prefix notation and CIDR notation explained. Finally, discussed the various fields of the IPv4 header.

---

[*] Email: dhanapal.ang@gmail.com.

Keywords: IP addressing, unicast, multicast, broadcast, IPv4 address classification, subnet, CIDR notation, IPv4 header

INTRODUCTION

Every computer in the network has to be identified uniquely, in order to communicate with each other, as like our home address. The computers in the networks assigned with Internet Protocol address. The Internet Protocol is part of the networking layer in the TCP/IP protocol suite. The Internet Protocol(IP) is responsible for addressing the computer in the network. The networking layer takes care of routing the packets from one computer to another in the computer network. Currently, we have two version of Internet Protocol exist in the networking. They are IP address version 4 (IPv4) and IP address version 6 (IPv6). This chapter discusses the details about IPv4 addressing schemes, various classifications, IPv4 header and associated basic concepts like Host, Network, Unicast, Multicast, Broadcast, etc.

IPV4 ADDRESSING

In the IPv4 addressing schemes, every device in the network represented with an IP address. The IPv4 address represented with 32bits (or) 4 bytes of decimal numbers, separated by period or dot. The example of an IP address is 192.168.12.34. The IP address is unique in the network so that the device identified uniquely in the network. Each IP address contains two portions. The first portion is the host address, and second is the network portion. The network portion identifies the particular network of the given IP address, and host part helps to locate the device within the given network.

The 32 bits IP address divided into four octets. Each octet in the IP address can take values ranges from 2^8. It means that every octet can have decimal values between 0 to 255. Let take an example of an IP address 192.168.12.34.

Dotted Decimal Notation to Binary Representation

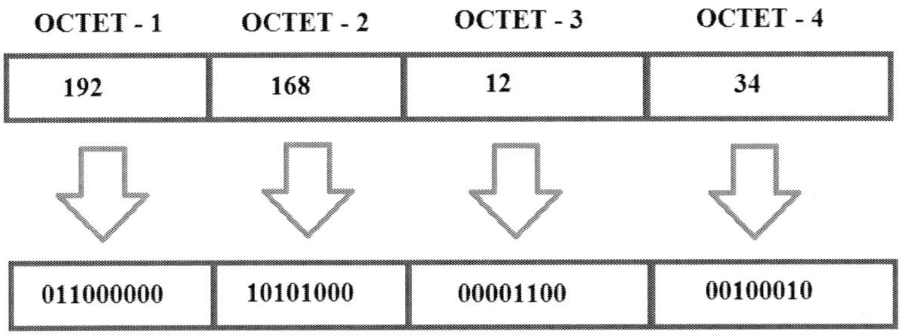

In the IPv4 address, the highest octet generally represents the network portion, and lower octet indicates the host portion. The IPv4 addressing schemes divided into multiple classes, based on the class type, the number of bits for network portion, and the host portions varies.

IPv4 Addressing Schemes

The IPv4 addressing schemes support multiple addressing modes. The behaviour of the device differs in the different addressing modes. The various addressing modes are

- Unicast Addressing
- Broadcast Addressing
- Multicast Addressing

The behaviour of the router for the each of the addressing modes explained as described below.

Unicast Addressing

In this mode, the exact device IP address used. This means that the IP address indicates the exact IP address of the destination device. The packets sent to the corresponding device.

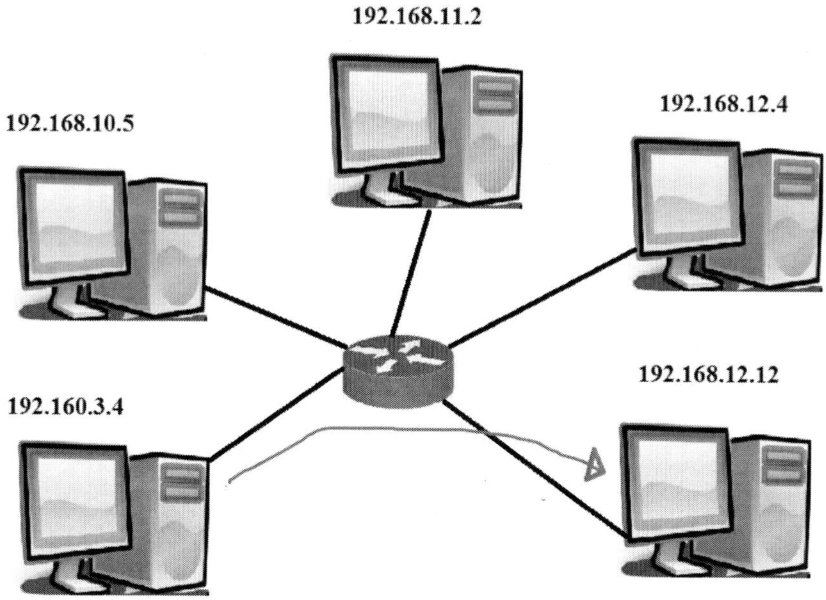

Figure 1. Unicast Addressing Mode.

In the given Figure 1, the device with an IP address 192.160.3.4 sends the packet to the system with an IP address 192.168.12.12. The packets sent only to the particular device which has the given destination IP address.

Broadcast Addressing

In this mode, the destination IP address shall have one or more octet as 255. This means that every node in the given network receives the packets except the device sending which sends the message.

Figure 2 represents the broadcasting addressing scheme where the device with an IP address 192.160.3.4 sends out a broadcast message to the IP address 192.255.255.255. The corresponding messages delivered to all the system attached to the network 192.X.X.X.

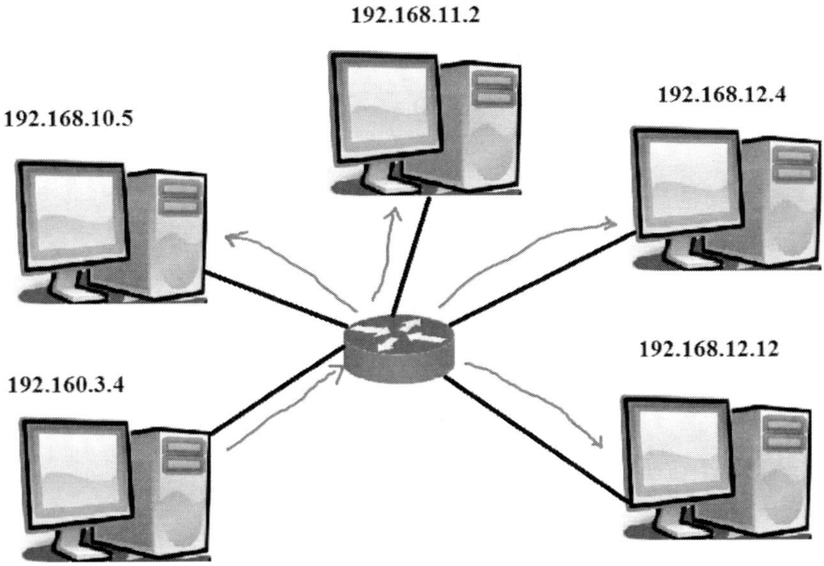

Figure 2. Broadcast Addressing Mode.

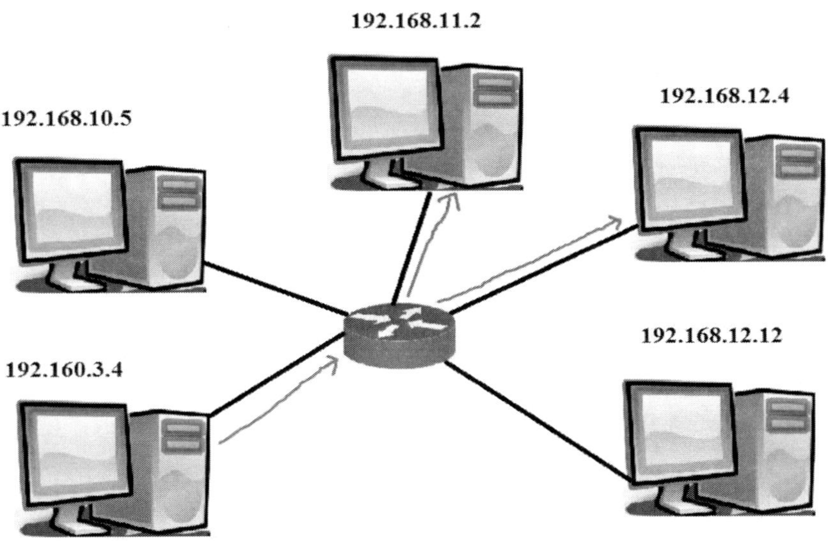

Figure 3. Multicast Addressing Mode.

Multicast Addressing

In this mode of addressing, the destination IP address looks like 224.X.X.X and the packets sent to this IP address will be delivered to the set of the host who is part of the given multicast group. The multicast addressing mode shown in Figure 3.

The device with an IP address 192.168.11.2 and 192.168.12.4 are part of the multicast group such as 224.0.1.2. Any packet sends to the destination IP address 224.0.1.2, delivered to IP address 192.168.11.2 as well as 192.168.12.4. It means that all the members of the multicast group are receiving the packets destined to the multicast address.

IPv4 Address Classifcations

There are five classifications of IPv4 address

- Class A Addresses
- Class B Addresses
- Class C Addresses
- Class D Addresses
- Class E Addresses

Class A Addresses

In this classification, the first octet represents the NETWORK ID, and octet 2 to 4 represents HOST ID as shown in Figure 3.

OCTET - 1	OCTET - 2	OCTET - 3	OCTET - 4
NETWORK ID	HOST ID	HOST ID	HOST ID

Figure 3. Class A IP address.

0	X	X	X	X	X	X	X

Figure 4. First octet of Class A address.

The first octet first bit is always set to 0 and looks like in Figure 4.

The possible values for this octet are 20 to 26. It takes values ranges from 1 to 127. We can have the maximum of 28 networks in Class A address and the maximum 224 hosts.

The total of 127 networks in class A, the IP address starting with 127 reserved for loopback address. So, the total available networks in the class A address is 126 networks and values ranging from 1 to 126.

Similarly, the host address ending with 0 used for representing the network address rather than host address. For example, the IP address 10.16.124.0 represents the class A network 10.X.X.X, but not the particular host in the network.

The IP address with last octet 255 used for the broadcast address of that particular network. For example, the IP address 10.12.124.255 used to send the broadcast message to all the system in the network 10.12.124.X. So the available number of host in class A is $2^{24} - 2 = 16777214$.

Class A addresses range from 1.X.X.X to 126.X.X.X.

Class B Addresses

In this classification, the octet-1 and octet-2 represent the NETWORK ID and octet 3 to 4 represent HOST ID as shown in Figure 5.

OCTET - 1	OCTET - 2	OCTET - 3	OCTET - 4
NETWORK ID	NETWORK ID	HOST ID	HOST ID

Figure 5. Class B IP address.

1	0	X	X	X	X	X	X

Figure 6. First octet of Class B Address.

The first octet first two bit are always set to 1 0 and looks like.

The possible values for this network identifiers are 20 to 214. The class B IP address accommodates the 16384 networks and 2^{16} - 2 (65534) hosts.

Class B addresses ranges from 128.X.X.X to 191.255.X.X.

Class C Addresses

In this classification, the octet-1 to octet-3 represents the NETWORK ID and octet-4 to represent HOST ID as shown in Figure 7.

The first octet first three bit are always set to 1 1 0 and looks as shown in Figure 8.

The possible network addresses in class C are 2097152 (2^{21}). The maximum number of host in each network are 254 (2^8-2). Class C addresses range from 192.X.X.X to 223.255.255.X.

OCTET - 1	OCTET - 2	OCTET - 3	OCTET - 4
NETWORK ID	NETWORK ID	NETWORK ID	HOST ID

Figure 7. Class C IP address.

1	1	0	X	X	X	X	X

Figure 8. First octet of Class C Address.

Class D Addresses

The Class D address first octet four bits are set to 1 1 1 0 as shown in Figure 9. This Class D IP addresses reserved for multicasting.

The range of IP used in this classification is 224.0.0.0 to 239.255.255.255.

1	1	1	0	X	X	X	X

Figure 9. First octet of Class D Address.

Class E Addresses

The Class E address is an experimental address and future use. The first four bits of the first octet set to 1 1 1 1 as shown in Figure 10.

1	1	1	1	X	X	X	X

Figure 9. First octet of Class E Address.

Subnetting and Subnet Mask

The IP address classes help to accommodate the IP networks by dividing the 32-bit IP address into two part such as NETWORK ID and HOST ID. Let's take the class A network, where it has the possibility of 16 million hosts on the same network. All the hosts on the same physical network bounded by IP routers share the same broadcast traffic, and they are in the same broadcast domain. Practically is not feasible to have 16 million system in the same broadcast domain. This results in many of the 16 million host addresses are unused and wasted. Even in the class B network having the 65K system in the same network is impractical.

In order to effectively use all possible IP address and create smaller broadcast domains, IP networks subdivided into smaller networks known as subnetworks. Each subnetwork belongs to one broadcast domain associated with an IP router, and every subnetwork bounded with its own subnetted network identifier. These subnets are part of the original class-based network. The subnetwork created by means of borrowing some of the bits from the associated HOST ID.

Each class of IP address, we have well-known NETWORK ID and HOST ID. In the case of subnetting, some of the higher order bits of the HOST ID used for subnetting, there arises a question of how to differentiate the SUBNET ID and HOST ID. In order to find out the SUBNET ID and HOST ID, there should be a corresponding subnet mask associated with it. This subnet mask helps to find out a number of bits used in the HOST ID to create the subnet with the given network.

The subnet mask is again 32-bit number looks similar to the IP address which tells which has all bits set to 1 for NETWORK ID and 0 for HOST ID. The default subnet mask for various classification discussed as follows.

Sometimes the subnet mask indicated using network prefix notation which is in the form of an IP address dotted decimal format/<number of bits>.

For example, the Class B IP address 135.56.23.53 with subnet mask 255.255.0.0 represented as 135.56.23.53/16 where /16 says that a number of bits used for network starting from octet-1. The default prefix notation for different IP address classes shown in Table 2.

Table 1. Default subnet masks

Class of Address	Default subnet mask in Dotted Decimal Format	Binary Representation of subnet mask
Class A	255.0.0.0	11111111 00000000 00000000 00000000
Class B	255.255.0.0	11111111 11111111 00000000 00000000
Class C	255.255.255.0	11111111 11111111 11111111 00000000

Table 2. Default Prefix Notation for IP address classes

Class of Address	Dotted Decimal Notation	Prefix Notation
Class A	255.0.0.0	/8
Class B	255.255.0.0	/16
Class C	255.255.255.0	/24

CLASSLESS INTER-DOMAIN ROUTING OVERVIEW

We have discussed various classes of IPv4 addressing so far. This is also known as classful addressing. The classifications impose the constraints that limiting the number of hosts and the network in the given class. The subnet mask helps to subside the issue but has the limitation of defining the fixed length subnets. This fixed length subnet leads to exhaustion of the IP address. To avoid faster exhaustion of the IPv4 address and slow down the growth of routing table to have routing entry on the router, The Internet Engineering Task Force (IETF) introduce the classless inter-domain routing(CIDR) in the year of 1993.

The CIDR based on the variable-length subnet masking (VLSM) method, and provide the flexibility of having the variable length prefixes. The CIDR notation is same as the prefix notation like 192.168.2.6/24. This indicates that 192.168.2.0 is the network and 192.168.2.6 is the host address. The CIDR introduced the concept of allocating the address blocks to the organizations based on their requirements. The collection of such multiple contiguous prefixes form the supernets on the Internet, and whenever possible advertised as aggregates, thus reducing the number of entries in the global routing table. The important difference between the subnet mask and the CIDR notation is, the previous one allowed to have non-contiguous bits in the prefix, but the CIDR always require contiguous prefix bits to form the aggregate network and reduce the route table entries. These aggregated groups known as the CIDR blocks. The number of IP address in the given prefix length calculated using the formula $2^{(\text{IP address Length} - \text{Prefix Length})}$ where IP address length in IPv4 is 32.

Table 3. IPv4 CIDR Table

Address with Prefix	The last IP address	Mask	Total Number of IP addressess
A.B.C.D/32	0.0.0.0	255.255.255.255	1
A.B.C.D/31	0.0.0.1	255.255.255.254	2
A.B.C.D/30	0.0.0.3	255.255.255.252	4
A.B.C.D/29	0.0.0.7	255.255.255.248	8
A.B.C.D/28	0.0.0.15	255.255.255.240	16
A.B.C.D/27	0.0.0.31	255.255.255.224	32
A.B.C.D/26	0.0.0.63	255.255.255.192	64
A.B.C.D/25	0.0.0.127	255.255.255.128	128
A.B.C.D/24	0.0.0.255	255.255.255.0	256
A.B.C.D/23	0.0.1.255	255.255.254.0	512
A.B.C.D/22	0.0.3.255	255.255.252.0	1,024
A.B.C.D/21	0.0.7.255	255.255.248.0	2,048
A.B.C.D/20	0.0.15.255	255.255.240.0	4,096
A.B.C.D/19	0.0.31.255	255.255.224.0	8,192
A.B.C.D/18	0.0.63.255	255.255.192.0	16,384
A.B.C.D/17	0.0.127.255	255.255.128.0	32,768
A.B.C.D/16	0.0.255.255	255.255.0.0	65,536
A.B.C.D/15	0.1.255.255	255.254.0.0	131,072
A.B.C.D/14	0.3.255.255	255.252.0.0	262,144
A.B.C.D/13	0.7.255.255	255.248.0.0	524,288
A.B.C.D/12	0.15.255.255	255.240.0.0	1,048,576
A.B.C.D/11	0.31.255.255	255.224.0.0	2,097,152
A.B.C.D/10	0.63.255.255	255.192.0.0	4,194,304
A.B.C.D/9	0.127.255.255	255.128.0.0	8,388,608
A.B.C.D/8	0.255.255.255	255.0.0.0	16,777,216
A.B.C.D/7	1.255.255.255	254.0.0.0	33,554,432
A.B.C.D/6	3.255.255.255	252.0.0.0	67,108,864
A.B.C.D/5	7.255.255.255	248.0.0.0	134,217,728
A.B.C.D/4	15.255.255.255	240.0.0.0	268,435,456
A.B.C.D/3	31.255.255.255	224.0.0.0	536,870,912
A.B.C.D/2	63.255.255.255	192.0.0.0	1,073,741,824
A.B.C.D/1	127.255.255.255	128.0.0.0	2,147,483,648
A.B.C.D/0	255.255.255.255	0.0.0.0	4,294,967,296

For example, given the prefix length of /27, the number of possible IPv4 addresses are $2^{32-27} = 32$. Table 3. Gives the details on IPv4 CIDR notation.

IPv4 Header Format Overview

The IPv4 header contains multiple information to carry the packet over IP Networks. The size of the IPv4 header is 20 bytes. This IPv4 header also contains some additional optional field in addition to the 20 bytes of information. The pictorial view of the IP header captured in Figure 10.

The IP header consists of 5 words each of the word consists of 4 bytes (32 bits). Figure 10 top row indicates four bytes starting with 0 to 3. The next row specifies the bit position in each word and starting with 0 to 31, totally 32bits.

<table>
<tr><th rowspan="2">Word</th><th colspan="8">0</th><th colspan="8">1</th><th colspan="8">2</th><th colspan="8">3</th></tr>
<tr><th>0</th><th>1</th><th>2</th><th>3</th><th>4</th><th>5</th><th>6</th><th>7</th><th>8</th><th>9</th><th>10</th><th>11</th><th>12</th><th>13</th><th>14</th><th>15</th><th>16</th><th>17</th><th>18</th><th>19</th><th>20</th><th>21</th><th>22</th><th>23</th><th>24</th><th>25</th><th>26</th><th>27</th><th>28</th><th>29</th><th>30</th><th>31</th></tr>
<tr><td>1</td><td colspan="4">Version</td><td colspan="4">IHL</td><td colspan="8">ToS</td><td colspan="16">Total Length</td></tr>
<tr><td>2</td><td colspan="16">Identification</td><td colspan="3">Flags</td><td colspan="13">Fragment Offset</td></tr>
<tr><td>3</td><td colspan="8">Time To Live</td><td colspan="8">Protocol</td><td colspan="16">Header Checksum</td></tr>
<tr><td>4</td><td colspan="32">Source IP Address</td></tr>
<tr><td>5</td><td colspan="32">Destination IP Address</td></tr>
<tr><td>6</td><td colspan="32">Options</td></tr>
</table>

Figure 10. IPv4 Header details.

The details on the different IP header fields are.

Version (Bit 0 – 3 of word 1): 4 Bits

The first bits 0-4 in the first word says that the version of the IP protocol. This may be 4 for IPv4 and 6 for IPv6.

IHL (Bit 4 – 7 of Word 1): 4 Bits

This is expanded as Internet Header Length and indicates the IP header length as 4 bytes word. This is usually having the value 5 indicating that IP header size of 20 bytes. The value may be greater than five in case if there are any option fields.

ToS (Bit 8 – 15 of Word 1): 8 Bits

Type of Service used for the purpose of indicating the quality of service required for the traffic. The real-time traffic like voice interaction over the internet, video on demand is high sensitive traffic which requires the higher quality of services. Such type of applications uses this field.

Total Length (Bit 16 – 31 of Word 1): 16 Bits

It is the length of the packet size measured in terms of octets. This includes IP header and data. The maximum length of the data is 65535 octets.

Identification (Bit 0 – 15 of Word 2): 16 Bits

This is an identifier added by the sender to reassemble the fragmented packet at receiver.

Flags (Bit 16 – 18 of Word 2): 3 Bits

These three bits used for setting various control flags for fragmentation.

- Bit 0: reserved, must be zero. (Most Significant Bit)
- Bit 1: (Dont Fragment Bit) 0 = May Fragment, 1 = Don't Fragment.
- Bit 2: (More Fragment Bit) 0 = Last Fragment, 1 = More Fragments.

Fragment Offset (Bit 19 – 31 of Word 2): 13 Bits

This field specifies the offset of a particular fragment relative to the beginning of the original IP datagram. The first fragment starts with 0.

Time to Live (TTL) (Bit 0 – 7 of Word 3): 8 Bits

The maximum amount of time the IP datagram allowed to live in the networks. The value 0 in this field says that datagram to be destroyed. The time measured in units of seconds.

Protocol (Bit 8 – 15 of Word 3): 8 Bits

This field indicates the next layer protocol involved in processing the data in the IP datagram.

Header Checksum (Bit 16 – 31 of Word 3): 16 Bits

This is a checksum used for verifying IP header field only. In each of the hop, this checksum verified for detecting the error. The checksums computed again and updated in the header due to the reason that TTL value gets updated before sending it out to next hop.

Source Address (Bit 0 – 31 of Word 4): 32 Bits

The source IP address.

Destination Address (Bit 0 – 31 of Word 5): 32 Bits

The destination IP address.

Options: Variable in Length

This is an optional field and vary in size.

SUMMARY

This chapter discussed the details on IPv4 addressing, the basic concepts like Host ID, Network ID, Unicast addressing, Broadcast addressing, Multicast addressing. Also, covered the classifications of IPv4 addressing, the concept of subnetting, need for subnetting, subnet mask. The prefix notation representation, CIDR notation and various fields of the IPv4 headers explained.

REFERENCES

Beasley, Jeffrey S. and Piyasat Nilkaew, *A Practical Guide to Advanced Networking*. Third Edition. Indiana, USA: Pearson Education; 2013.

Dean, Tamara, *Network+ Guide to Networks*. Sixth Edition. USA: Cengage Learning; 2013.

Highteck (2017). *Addressing the Network - IPv4*. [Online] Accessed on: [2017-12-01]. URL: http://www.highteck.net/EN/Network/Addressing_the_Network-IPv4.html.

Microsoft (2017). *Address Classes*. [Online] Accessed on: [2017-12-01]. URL: https://technet.microsoft.com/en-us/library/cc940018.aspx.

Microsoft (2017). *Subnets and Subnet Masks*. [Online] Accessed on: [2017-12-01]. URL: https://technet.microsoft.com/en-us/library/cc958832.aspx.

Tanenbaum, Andrew S. and David J. Wetherall, *Computer Networks*. Fifth Edition. US: Prentice Hall; 2011.

Tutorialspoint (2017). IPv4 Addressing. [Online] Accessed on: [2017-12-01]. URL: https://www.tutorialspoint.com/ipv4/ipv4_addressing.htm.

Wikipedia (2017). IP address. [Online] Accessed on: [2017-12-01]. URL: https://en.wikipedia.org/wiki/IP_address.

Wikipedia (2017). IPv4. [Online] Accessed on: [2017-12-01]. URL: https://en.wikipedia.org/wiki/IPv4.

Wikipedia (2017). Multicast address. [Online] Accessed on: [2017-12-01]. URL: https://en.wikipedia.org/wiki/Multicast_address.

BIOGRAPHICAL SKETCH

A. Dhanapal

Affiliation: VIT University, Chennai, India

Education: BE, (CSE) and ME, (CSE)

Research and Professional Experience: Currently perusing the research at VIT University in Chennai, Tamil Nadu, India. The area of research includes security in the cloud computing.

A. Dhanapal received his BE degree in Computer Science and Engineering from Bharathiar University, India. He received his ME in Computer Science and Engineering from Sathyabama University, India. He is pursuing PhD in the field of Cloud Computing Security at VIT University, India. He has 10+ years of industry experience in the various areas like high-performance computing, storage and networking. His area of interests includes Cloud Computing, Cloud Security, Networking, Virtualization, etc.

V. Vijayakumar, PhD

Affiliation: Professor & Associate Dean, School of Computing Science and Engineering, VIT University, Chennai Campus, India
Email: vijayakumar.v@vit.ac.in

Vijayakumar Varadarajan is currently a Professor and an Associate Dean for School of Computing Science and Engineering at VIT University, Chennai, India. He has more than 18 years of experience including industrial and institutional. He also served as a Team Lead in industries like Satyam, Mahindra Satyam and Tech Mahindra for several years. He has completed Diploma with First Class Honors. He has completed BE CSE and MBA HRD with First Class. He has also completed ME CSE with First Rank Award. He has completed his PhD from Anna University in 2012. He has published many articles in national and international level journals/conferences/books. He is a reviewer in IEEE Transactions, Inderscience and Springer Journals. He has initiated a number of international research collaborations with universities in Europe, Australia, Africa, Malaysia, Singapore and North America. He had also initiated joint research collaboration between VIT University and various industries. He is also the Guest Editor for few journals in Inderscience, Springer and IGI Global. He also organized several international conferences and special sessions in USA, Vietnam, Africa, Malaysia and India including IEEE, ACSAT, ISRC, ISBCC, ICBCC etc. His research interests include computational areas covering grid computing, cloud computing, computer networks, cyber security and big data. He received his university-level Best Faculty Award for 2015–2016. He is also a member of several national and international professional bodies including EAI, BIS, ISTE, IAENG, CSTA, IEA etc.

In: Network Theory and Analysis
Editor: Amber Jorgensen
ISBN: 978-1-53613-460-5

Chapter 5

AN OVERVIEW OF THE IPV6 ADDRESSING SCHEME

***A. Dhanapal*[1,*] *and P. Nithyanandam*[2]**
[1]School of Computing Science & Engineering,
VIT University, Chennai, India
[2]School of Computing Science and Engineering,
VIT University, Chennai Campus, Chennai, India

ABSTRACT

This chapter gives the introduction of the IPv6 addressing and the need for IPv6 addressing. This further discusses the IPv6 Addressing formats, the different address types, the special addresses in the IPv6. Finally covers the various fields of the IPv6 and the difference between the IPv4 and the IPv6 addressing scheme.

---

* Correspondence: School of Computing Science & Engineering, VIT University, Chennai, India. Email: dhanapal.ang@gmail.com.

Keywords: IPv6 addressing, IPv6 address format, IPv6 address types, IPv6 special addresses, IPv6 header and comparision of IPv4 and IPv6 addressing

INTRODUCTION

Internet Protocol Version 6 or in short IPv6 is a successor of Internet Protocol Version 4 or IPv4 address. The IPv6 addressing designed to solve the inherent problem the IPv4 address exhaustion. Since the IPv4 addressing uses 32 bits to represent an IP address, which can accommodate the maximum of 2^{32} IP addresses. The IPv6 uses 128 bits to represent an IP address so comes up with enormous scale of the IP addresses (2^{128}) when compared to IPv4. The original intention of the IPv6 addressing scheme is to replace the IPv4 addressing scheme, but due to the introduction of Classless Inter-Domain Routing or CIDR allocation within the IPv4 addressing, the IPv4 continues to exist in a prolonged span. The participation of the various devices like mobile phones, smart electronic devices, etc., IPv4 address get exhausted soon. As per Google Statistics on Dec 2017, 21.85% Internet traffic uses IPv6 addresses and remaining all still running with the IPv4 addresses.

IPV6 ADDRESSING

The latest version of Internet Protocol is IPv6. Due to the rapid growth of the device participation on the Internet, cannot be accommodated with 32bit length IPv4 addresses. The exhaustion of IPv4 addresses is arriving at a faster rate than predicted. In order to solve this issue, the IPv6 introduced by Internet Engineering Task Force (IETF). The IPv6 captured in RFC 2460, Internet Protocol and Version 6 (IPv6) specification by IETF initially. The IPv6 header format has been modified to describe its architecture and supported services. The IPv6 has been designed to make existing IPv4 users migrate easily to IPv6 while providing services like Quality of Service

(QoS), end-to-end security, unique addresses, etc. The IPv6 address based on the prefix notation and it aggregates networks efficiently as discussed in the CIDR notation and reduces the number of routing entries on the router. The IPv6 supports globally deployed routing protocols such as Open Shortest Path First (OSPF), Border Gateway Protocol (BGP), Routing Information Protocol (RIP), etc.

The IPv6 address represented using 128-bits or quadruples of the 32-bit address. It means IPv6 address length is four times of the IPv4 address and accommodates the much much higher number of unique addresses.

IPv6 Address Formats

The IPv6 addresses represented as the collection of 16-bit hexadecimal identifiers separated by colons (:) and looks as follows:

XX:XX:XX:XX:XX:XX:XX:XX.

The example IPv6 address are:

- 1002:DB8:764:310:FEDB:BA96:7654:3210
- 2000:DB8:10:0:400:1800:200C:447A

The IPv6 addresses may contain successive zeros in the fields. Double colons (::) used to compress such successive zeros of the hexadecimal fields at the beginning, middle, or end of an IPv6 address. The user may configure more than one IPv6 address for the single interface but allows to configure only one link-local address. Table 1 gives example for the standard format as well as the compressed format of the IPv6 address.

The double colons (::) can only be used once in the IPv6 address to compress the longest successive zeros of the hexadecimal fields. The IPv6 address hexadecimal letters are case insensitive.

The IPv6 address "0:0:0:0:0:0:0:1" or "::1" is the loopback address used for sending packets to itself. This is same as the loopback address

"127.0.0.1" in the IPv4 address. The important note is that the IPv6 loopback address cannot be assigned to any physical interface. Any IPv6 device that does not forward packets can have loopback address as their source or destination address.

Table 1. IPv6 standard and compressed formats

Zero's Location	IPv6 Standard Format	IPv6 Compressed Format
Continous 0's near begining	2000:0:0:0:CB8:800:200D:417A	2000::CB8:800:200D:417A
Continues 0's in the middle	FF05:0:0:0:0:0:0:101	FF05::101
Starting with continuous 0's	0:0:0:0:0:0:0:1	::1
Ending with continuous 0's	0:0:0:0:0:0:0:0	::

Similarly, the address with all zeros "0:0:0:0:0:0:0:0" or "::" in table 1 is known as the unspecified address. The newly introduced system or node in the IPv6 network usually have the unspecified address until it assigned with an IPv6 address. As like IPv6 loopback address, the unspecified IPv6 address cannot be assigned to any interface. This unspecified address should not be used as destination addresses in IPv6 packets. The prefix notation representation of an IPv6 address looks like

XX:XX:XX:XX:XX:XX:XX:XX/Prefix_length.

where Prefix_length is a decimal value represents a number of consecutive 1's in the network address. An example of an IPv6 address in prefix notation is 2000:DB8:8086:6502::/32. The decimal value 32 indicates first 32 bits are 1's used to represent the network addresses.

IPv6 Address Types

The IPv6 addresses are classified into the following types and those types can be further categorized by their type and scope.

- Unicast addresses
- Multicast addresses
- Anycast addresses.

Note: There is no Broadcast address in the IPv6, and Anycast address is the new type available in IPv6.

Unicast Addressing

This is same as IPv4 Unicast addressing mode and the exact IP address of the destination device used for communication. The packets sent to the corresponding device.

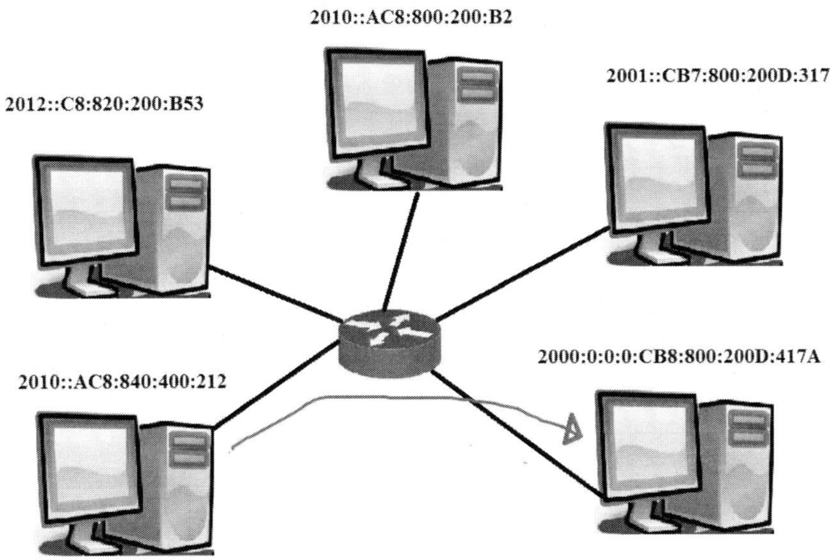

Figure 1. IPv6 Unicast Addressing.

In the given Figure 1, the device with an IPv6 address 2010::AC8:840:400:212 sends the packet to the system with the IPv6 address 2010::CB8:800:200D:417A. The packets received only by the particular device which has the given destination IPv6 address.

Multicast Addressing

This is also same as an IPv4 multicast addressing mode and shown as in Figure 2.

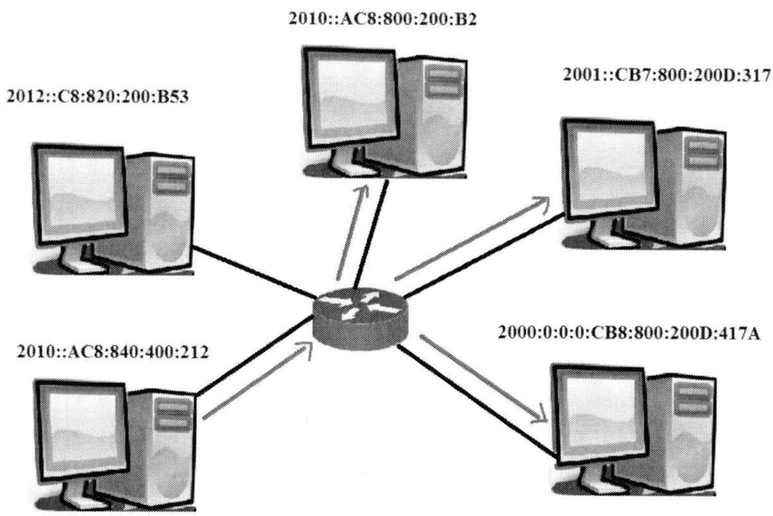

Figure 2. IPv6 Multicast Addressing.

The IPv6 devices registers for the particular IPv6 multicast group and whenever any device sends the packet to the multicast IPv6 address, all the subscribers of that multicast address will receive the packet.

In Figure 2. the device with IPv6 address 2010::AC8:840:400:212 sends packets to an IPv6 multicast address FF01:0:0:0:0:0:0:101. The subcribers to this IPv6 address are 2010::CB8:800:200D:417A, 2011::CB7:800:200D:317 and 2010::AC8:800:200:B2. The packets are destined to their subscribers.

Anycast Addressing

This is the new addressing type introduced in the IPv6 addressing where if any system in the IPv6 network sends the packet to the anycast IPv6 address, the system which has less routing distance and anycast IPv6 address will receive it. Figure 3 captures the working of anycast addressing.

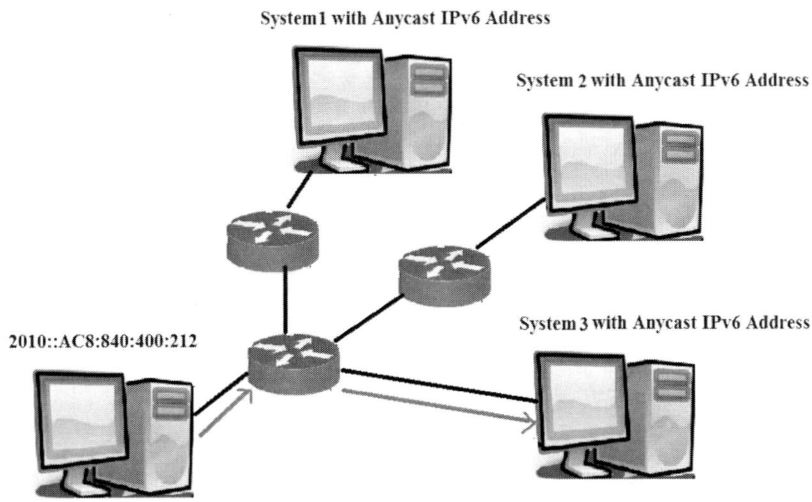

Figure 3. IPv6 Anycast Addressing.

In Figure 3, the system with an IPv6 address 2010::AC8:840:400:212 sends the packet to the anycast IPv6 address. The network has three routers and three systems with anycast IPv6 address. The system1 and system2 are having the distance of 2 from the source as they have to traverse through 2 hops. The system3 has the distance of 1 from the source as it is reachable via the only one router. In anycast addressing packet send to the system 3 as it has less distance when compared to other systems.

The Unicast and anycast IPv6 addresses can further be classified based on the scope as follows

- *Link-local:* This is a local IPv6 address used by system communicate within the subnet. The link-local addresses always start with 1111111010 (FE80). In the IPv6 network, the interface always receives a link-local address even if the user assigns the IPv6 address manually or an interface receives the IPv6 address via DHCP server. This indicates that the computers on a network or link can always communicate using the IPv6 link-local addresses. The router never forwards the link-local addresses, so it can only be reached via the link. This is the reason for next 54 bits set to 0 in the link-local address. The operating system randomly assigns the last 64 bits of the link-local address.

- *Site-local:* In IPv6, the site local address reserved for organization and packets routed within this private space. The packets cannot be routed to the Internet. The first 10 bits of a site-local address should be 1111111011. This is the reason for site-local addresses are always starts with FEC0. The next 54 bits used to represent the subnet ID, which administrator can use in the organization for hierarchical routing. The final 64 bits indicates the interface ID and those 64 bits should uniquely represent the host within the organization. So, the prefix of the site-local address is FEC0::/10. The important note is that this site-local address deprecated and whoever uses can continue to exist for some more time.
- *Global:* These IPv6 addresses are globally routable and communicated with the Internet.

IPv6 Special Addresses

The IPv6 defines various special addresses, listed as

- Unspecified Address.
- Loopback Address.
- Default Route.
- Reserved Multicast Address for Routing Protocols.
- Reserved Multicast Address for Routers/Node.
- IPv4 Mapped Address.
- 6to4 Addresses.
- Solicited-Node Address.

The details on the special addresses discussed as follows

Unspecified Address

As we have discussed earlier, unspecified address is "0:0:0:0:0:0:0:0." Initially when node coming up in the IPv6 network has this unspecified address. Until IPv6 address to the node, this address remains be there. The compressed format of this address is "::" (double colon). The corresponding prefix notation is ::/128.

Loopback Address

The loopback address in the IPv6 address specified as "0:0:0:0:0:0:0:1" or "::1". The packet with loopback address never leaves the node. The prefix notation for the loopback address is ::1/128.

Default Route

The default route in the IPv6 address represented as "0:0:0:0:0:0:0:0" or simply "::". This default route used as the entry in the routing table when there are no specific route matches for the given destination, the router forwards the packet using the default route. The prefix notation for default route is ::/0.

Reserved Multicast Address for Routing Protocols

There are certain IPv6 multicast addresses reserved for the specific use of routing protocols. The IPv6 multicast address and corresponding routing protocols captured in table 2.

Table 2. IPv6 Reserved multicast address for routing protocols

IPv6 Address	The Corresponding Routing Procol
FF02:: 5	OSPFv3
FF02:: 6	OSPFv3 Designated Routers
FF02:: 9	RIP
FF02:: A	EIGRP

Reserved Multicast Address for Routers/Node

There are some multicast IPv6 addresses reserved for routers or node. These addresses help routers or node to talk to available routers or nodes on a network segment with configured with the IPv6 address. The list of addresses shown in table 3.

Table 3. Reserved Multicast addresses for Routers and Node

IPv6 Address	The scope
FF01:: 1	All nodes in the interface local
FF01:: 2	All routers in the interface local
FF02:: 1	All nodes in the link-local
FF02:: 2	All routers in the link-local
FF05::2	All routers in the site-local

IPv4 Mapped Address

The IPv4 mapped addresses used only for internal representation on a host. This allows the programmer to use the single API for both the IPv4 address as well as the IPv6 address. The format of the IPv4 mapped addresses is "::FFFF:X.Y.Z.W".

The IPv4 address "X.Y.Z.W" is mapped to IPv6 address as "::FFFF:X.Y.Z.W".

6to4 Addresses

6to4 is a tunnelling technique that allows two IPv6 networks to communicate over an IPv4 network. This is useful especially in case of the Internet. In order to happen this tunnelling to work at least one of the IPv6 node needs to have an IPv4 address and a 6to4 special IPv6 address. A system which has a 6to4 address known as "6to4 host." The 6to4 addresses have the prefix 2002::/16. The followed 32 bits represent the IPv4 address of the 6to4 host converted into hex. Since a 6to4 address shall only be created by means of using a unique public IPv4 address, the equivalent IPv6 address is unique. The final 80 bits used for the local network and the host.

Solicited-Node Address

The special type of multicast address is the solicited-node addresses. This prefix notation of the solicited address looks like FF02::1:FF00:0/104. The solicited-node addresses used for neighbour discovery. This is same as ARP in the IPv4, like when a host wanted to know the MAC address for certain IPv6 address, then it would use solicited multicast address to contact only those hosts that listen to the solicited-node addresses.

AN OVERVIEW OF IPV6 HEADER FORMAT

An IPv6 data packet consists of two parts

- The IPv6 header.
- The payload.

The header is part of the first 40 bytes. Figure 4 describes the IPv6 packet header.

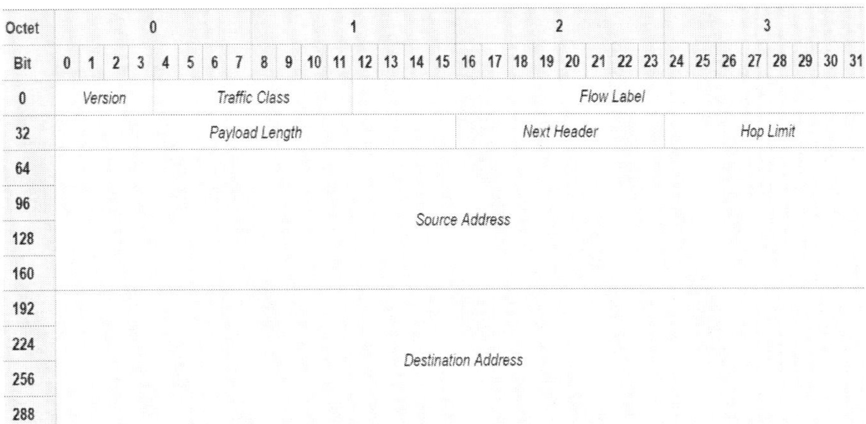

Figure 4. IPv6 Header Format.

The details on the different IP header fields are

Version (Bit 0 – 3 of word 1): 4 Bits

The 0-3 bits in the first word says that the version of the IP protocol. The value is 6 for the IPv6 address.

Traffic Class (Bit 4 – 11 of word 1): 8 Bits

The 8 bits field used for providing priority to the packets so that the routing happens as per the value. This field represents the Quality of Service (QoS) provided to the packet.

Flow Label (Bit 12 – 31 of word 1): 20 Bits

This field used for the purpose of giving special services to the real-time applications. When this set to a non-zero value, it hints to routers/switches with multiple outgoing paths that these packets should take the same path, in order to avoid the reordering.

Payload Length (Bit 0 – 15 of word 2): 16 Bits

It is the length of the packet size measured in terms of octets. This includes any extension headers. During when an hop-by-hop extension header carries the jumbo payload option, this field set to 0.

Next Header (Bit 16 – 23 of word 2): 8 Bits

This indicates the type of header that immediately follows the IPv6 header. It has the same values as the IPv4 protocol field.

Hop Limit (Bit 24 – 31 of word 2): 8 Bits

This holds the value of 8 bits unsigned integer value and decremented by one by each hop which forwards the packet. The packet discarded when this value becomes zero.

Source Address (word 3 to word 6): 128 bits

The source IPv6 Address.

Destination Address (word 7 to word 10): 128 bits

The destination IPv6 address.

The IPv6 Extension Headers

The IPv6 options can be kept in the separate extension headers and located between the IPv6 header and the transport-layer header of the packet. The IPv6 extension processed by the final destination rather than examined at any router until to reach the destination. This improves the performance of the router for the packets which comes with optional values in the header. The IPv6 extension headers currently defined are:

- Routing – Used for extended routing.
- Fragmentation – Used for the purpose of fragmentation and reassembly.
- Authentication – Used for integrity, authentication and security.
- Encapsulating Security Payload – For carrying the confidentiality details.
- Hop-by-Hop options – Special options that require hop-by-hop processing.
- Destination options – Optional information to be examined by the destination node.

The Comparison of an IPv4 and an IPv6 Addressing

Table 4 captures the key difference between the IPv4 addressing and the IPv6 addressing.

Table 4. The comparision of an IPv4 and an IPv6 addressing

The IPv4 Addressing	The IPv6 Addressing
Address length is 32 bits.	Address length is 128 bits.
Generally, the IPv4 addresses represented in decimal.	These addresses represented in Hexadecimal.
Broadcasting is available.	Broadcasting is not available.

Table 4. (Continued)

The IPv4 Addressing	The IPv6 Addressing
Anycasting not available.	Anycasting supported.
Address Resolution Protocol (ARP) used for map IPv4 address to MAC address.	ARP replaced with Neighbor Discovery Protocol (NDP).
Internet Group Management Protocol (IGMP) used to manage multicast group membership.	IGMP replaced with Multicast Listener Discovery (MLD) messages.
The checksum field is available in the header.	No checksum field available in the header.
The IP Security is an optional.	The inbuild IP Security support available.
No field to indicate the packet flow.	The flow label field used for packet flow identification.
Manual or DHCP configuration needed.	Auto configuration is available.

SUMMARY

This chapter discussed the details on the IPv6 addressing scheme and need for the IPv6 addressing. This further covered the IPv6 addressing formats, different address types, the special addresses in the IPv6. Finally touched upon the various fields of the IPv6 header, and the key difference between the IPv4 and the IPv6 addressing scheme.

REFERENCES

4sysops (2017). *IPv6 special addresses.* [Online] Accessed on: [2017-12-10].

Cisco (2017). *IPv6 Addressing and configuration guide.* [Online] Accessed on: [2017-12-10].

Dean, Tamara, *Network+ Guide to Networks*. Sixth Edition. USA: Cengage Learning; 2013.

Beasley, Jeffrey S. and Piyasat Nilkaew, *A Practical Guide to Advanced Networking*. Third Edition. Indiana, USA: Pearson Education; 2013.

Microsoft (2017). *IPv6 Address Types*. [Online] Accessed on: [2017-12-10].

Omnisecu (2017). *Difference between IPv4 and IPv6.* [Online] Accessed on: [2017-12-10].

Oracle (2017). *IPv6 Addressing Overview.* [Online] Accessed on: [2017-12-10].

Tanenbaum, Andrew S. and David J. Wetherall, *Computer Networks*. Fifth Edition. USA: Prentice Hall; 2011.

Tutorialspoint (2017). IPv6 Overview. [Online] Accessed on: [2017-12-10].

URL: http://www.omnisecu.com/tcpip/ipv6/differences-between-ipv4-and-ipv6.php.

URL: https://4sysops.com/archives/ipv6-tutorial-part-8-special-addresses/

URL: https://en.wikipedia.org/wiki/IPv6_address.

URL:https://www.cisco.com/c/en/us/td/docs/ios-xml/ios/ipv6_basic/configuration/xe-3s/ip6b-xe-3s-book.pdf.

URL: https://docs.oracle.com/cd/E23823_01/html/816-4554/ipv6-overview-10.html#scrolltoc.

URL: https://technet.microsoft.com/en-us/library/cc757359(v=ws.10).aspx

URL: https://www.tutorialspoint.com/ipv6/ipv6_overview.htm.

Wikipedia (2017). IPv6 Address. [Online] Accessed on: [2017-12-10].

Biographical Sketch

P. Nithyanandam, PhD

Affiliation: School of Computing Science and Engineering, VIT University, Chennai Campus, Chennai, India

Email: Nithyanandam.P@vit.ac.in

Dr. P. Nithyanandam is a Professor in School of Computing Science and Engineering, VIT University, Chennai Campus. He is an academician and has an experience of 15 years. He received BE in Computer Science and Engineering from Madurai Kamaraj University in 2000, MTech in Computer Science and Engineering from BITS pilani in 2003 and PhD in Computer Science and Engineering, from Anna University, Chennai in 2013. He has published many national and International publications to his credit and conducted various national and international conferences. He is also a reviewer for various leading journals such as IEEE, Elsevier, Inderscience, etc. He is the Head of Research Group 'Information Security' of VIT University Chennai campus. He has completed two funded projects for the Tamil Virtual University, funded by Tamilnadu Government. Currently taken an assignment in doing a consultancy work on the title "Image Theft Identification using Biometric". His research area include data security, malware analysis, compression, cloud computing and image processing. He is guiding six research scholars for PhD and MS program.

INDEX

D

E

F

G

H

I